WATER BALANCE OF MONSOON ASIA

WATER BALANCE OF MONSOON ASIA

—A Climatological Approach—

Edited by
Masatoshi M. YOSHINO

UNIVERSITY OF HAWAII PRESS
Honolulu 1971

Originally Published in 1971 by
UNIVERSITY OF TOKYO PRESS
Supported in part by the Ministry of Education
under the Publication Grant-in-Aid.

Library of Congress Catalog Card Number 78–168976
ISBN 0–87022–895–1

Printed in Japan

List of Contributors

AIHARA, Haruhiko   Department of Geography, Hosei University, Chiyodaku, Tokyo.

ASAKURA, Tadashi   Long-range Forecast Division, Japan Meteorological Agency, Chiyodaku, Tokyo.

HIRANUMA, Yoji   Forecast Division, Japan Meteorological Agency, Chiyodaku, Tokyo.

KAWAMURA, Takeshi   Forecast Division, Japan Meteorological Agency, Chiyodaku, Tokyo.

KAYANE, Isamu   Department of Geography, Tokyo University of Education, Bunkyoku, Tokyo.

KUBOTA, Isao   Niigata Local Meteorological Observatory, Niigata City, Niigata Prefecture.

KURASHIMA, Atsushi   Forecast Division, Japan Meteorological Agency, Chiyodaku, Tokyo.

MIZUKOSHI, Mitsuharu   Institute of Geography, Mie University, Tsu City, Mie Prefecture.

NEYAMA, Yoshiharu   Hiroshima Local Meteorological Observatory, Hiroshima City.

TABUCHI, Hiroshi   Department of Geography, Hosei University, Chiyodaku, Tokyo.

TSUCHIYA, Iwao   Meteorological Research Institute, Koenji-Kita, Suginamiku, Tokyo.

URUSHIBARA, Kazuko   Department of Geography, Hosei University, Chiyodaku, Tokyo.

WADA, Hideo   Long-range Forecast Division, Japan Meteorological Agency, Chiyodaku, Tokyo.

YOSHIMURA, Minoru   Institute of Geography, Yamanashi University, Kofu City, Yamanashi Prefecture.

YOSHINO, Masatoshi M.   Department of Geography, Hosei University, Chiyodaku, Tokyo.

Preface

In recent years, water balance has been one of the most important problems not only in the field of pure and applied climatology but also in the areas of meteorology, hydrology, and agricultural science and technology. Therefore, water balance has been the topic of studies by the International Hydrological Decade project. Generally, the conditions of water balance have been treated either on a global scale or, quite the opposite, from the hydrological viewpoint in specific drainage areas or from the standpoint of small-scale physical meteorology. For this reason, a need was felt for macro-scale climatological studies of water balance problems in regions such as Monsoon Asia, that is, studies that are narrower in scale than global studies, but wider than the studies of individual drainage areas.

Climatologically, Monsoon Asia is a very interesting region because the monsoons bring heavy, torrential rainfall with great interannual variation and striking differences from place to place. Furthermore, many relatively dry regions are unexpectedly found adjacent to humid regions. These characteristics of the area have been studied for many years, but they are not yet perfectly understood. Our goal, therefore, has been to clarify these phenomena in order to contribute to understanding water balance in Monsoon Asia.

The papers presented in this monograph describe group studies on water balance of Monsoon Asia supported by the Japanese Ministry of Education (Hydrology 1969—No. 91012). In addition, the ministry has also provided financial support for publication of the present book (1970–304). We would like to thank the ministry for this support, Mr. H. Tamiya for his assistance in proofreading, and the staff of the University of Tokyo Press for their kindness in arranging for publication.

Tokyo
March 1971

M. M. Yoshino
Editor

Contents

PART I

GENERAL INTRODUCTION

WATER BALANCE PROBLEMS IN MONSOON ASIA FROM THE VIEWPOINT OF CLIMATOLOGY

Masatoshi M. Yoshino

Abstract: This review paper deals with the problems of water balance in Monsoon Asia from the climatological standpoint. In particular, its intent is to present the historical background for the studies included in this volume. The following subjects have been considered: 1) Precipitation distribution during the monsoon seasons, with special regard to amount, number of rainy days, variability and distribution patterns in connection with synoptic patterns. 2) Climatic indices as an expression of water balance. 3) Recent results in synoptic climatological studies of summer circulation and the rainy season in early summer in East Asia. 4) The state of water vapor transport in Monsoon Asia, with special emphasis on the hydrological cycle in the upper Yangtze region and the mean total water transport field in Monsoon Asia. 5) A summary of climatic fluctuations in Monsoon Asia in geological and recent times. 6) A brief introduction to studies on water balance as related to water resources and water utilization.

1. INTRODUCTION

Water balance is one of the basic problems in modern climatology, but it has also been the center of discussion in classical climatology. The state of water balance as well as heat balance in a certain area or region represents its so-called climatic condition. This is always true, regardless of the definition of "climate."

Monsoon Asia, that is, South, Southeast and East Asia, is climatologically one of the most interesting regions in the world. Major focuses of interest include the structures of monsoons or seasonal changes in wind systems, rainfall characteristics associated with monsoons, "heat lows" at the surface level, and anticyclones in the higher troposphere over South Asia together with the tropical easterly jet in summer, subtropical anticyclones over the Northwest Pacific, the ITC and tropical cyclones, the huge and intense anticyclone over Siberia in winter and the Pacific polar frontal zones associated with the world-strongest subtropical westerly jet over East Asia in winter.

On the other hand, studies of water balance are one of the more important projects of the International Hydrological Decade. These, however, have so far been directed generally to specific drainage areas from the viewpoint of hydrology, to specific points from the viewpoint of micrometeorology or, quite the opposite, to the whole hemisphere by an analytic method of physical meteorology developed recently. For this reason, it was thought that climatological studies on water balance in Monsoon Asia, whose scale is wide but still less than global, must be of importance, since they have so far been neglected.

The present paper will review these problems and attempt to provide the background for the other studies in this monograph.

One product of these studies is a bibliography on water balance in Monsoon Asia published by Tsuchiya (1969). It is to be hoped that this will serve as a source of information concerning a number of subjects which are not mentioned in this paper.

2. Precipitation Distribution during Monsoons

A simple representation of the state of water balance is a precipitation map, because precipitation plays one of the most important roles in the water balance equation, varying much more from place to place than the other terms. Furthermore, precipitation in Monsoon Asia is strongly characterized by its seasonality, interannual variability, torrential behaviour and sharp regionality. These characteristics are far more intense than was previously anticipated.

Since the end of the last century, our knowledge of precipitation distribution in the world has become increasingly concrete. In this respect, there are classic descriptions on a global scale such as those by Woeikov (1887), Hann (1910), Kendrew (1927) and the more recent ones by Trewartha (1954, 1962) and Blüthgen (1966).

For Monsoon Asia, early treatises by geographers such as Passerat (1906) and Sion (1928) should be noted. Fuller descriptions from the standpoint of climatography before or during World War II were attempted by Arakawa (1942), Fukui (1942) and Ogasawara (1945). The latter analyzed climate on the basis of the concept of air mass theory or dynamic processes. He dealt with air masses, ITC, cyclogenesis and frontogenesis in Monsoon Asia and with regional and seasonal distributions of precipitation in relation to topographical conditions, applying the dynamic concept. He also presented a regional treatment of Monsoon Asia, dividing it into eight regions. More recent contributions are found in the climatography entitled "World Survey of Climatology" (Arakawa 1969).

Rainfall characteristics have also been analyzed, for instance, a work by the South Asia Institute, Heidelberg University (Schweinfurth et al., 1970), but few results have been published on humidity and evaporation, because of a scarcity of data. Even in the climatography by Ogasawara mentioned above, relative humidity and evaporation were only described briefly in classical terms with some additional details on diurnal change and the Föhn effect on local systems in several regions. Generally, it can be said, therefore, that little attention has been paid to water balance problems in Monsoon Asia.

After World War II, many books on the regional geography of Monsoon Asia were published. Lack of space forbids a complete discussion of all these works, but in general it can be said that descriptions of precipitation in Monsoon Asia became more detailed and concrete with modern techniques in synoptic and aerological climatology and with the aid of statistics. For instance, Johnson (1969) dealt with the mechanism of monsoons and ITCs, introducing results obtained through recent developments in synoptic meteorology and aerology, and discussed rainfall distribution in connection with types of seasonal change. He presented a map showing rainfall incidence over South Asia summarizing precipitation data in the sense of climatic dynamics. Ten types of rainfall incidence were distinguished on the basis of the occurrence of rainy months, which were defined as having more than one-twelfth of the mean annual rainfall. He also showed both monthly and annual rainfall dispersion diagrams. It is especially interesting to note how great is the range between the absolute minimum and the absolute maximum or between the lower quartile and the upper quartile of the annual amount of rainfall. According to Johnson's figures, Cox's Bazaar on the eastern side of the Bay of Bengal and Cochin on the west coast have a very great range together with a large amount of annual rainfall. Humidity and evaporation data at several stations in China and Korea (Watts, 1969) and in the Philippines (Flores et al., 1969) have been recently published.

The regional distributions of high and low values of monthly precipitation in June and July over East Asia have been given in a previous study (Yoshino, 1963). In that study, the high value was defined as the mean value of the upper decile of the monthly precipitation observed in a long period. The low value was defined in the same way as the mean value of the lower decile. South of Japan, the zone about 25°N in the Pacific has very small high and low values in June and also very small high values in July. This zone and its positional change coincide with the region of low water vapor transport. In contrast, large high and low values appear in the frontal zones and along the ITCs. Information on the monthly precipitation distribution in the oceans is still scanty. One study was made after World War II by Albrecht (1951), who devoted much effort to studying the water budget of the earth.

Rainfall distribution in relatively restricted regions such as India, Malaysia, China and Japan has been studied intensively. A number of papers can be found in Tsuchiya's bibliography (1969). The problems they cover can be summarized briefly as follows: 1) Local differences in the distribution and number of rainy days, such as studied by Domrös (1968, 1970) for India and its surroundings during the winter monsoon, pre-monsoon and summer monsoon seasons. 2) Rainfall variability studies, for example, for India by Naqvi (1949), Chatterjee (1953) and Sircar (1958), for Eastern China by Yao (1958, 1965) and for Japan by Daigo (1947). 3) The pattern of rainfall distribution during the monsoon over India such as presented by Anathakrishnan et al. (1964). 4) Wet and dry spells analyzed for West India by Raman et al. (1960), for China by Lu et al. (1965) and for Japan by Maejima (1967). 5) Heavy rainfall associated with cyclone or typhoon activities. This has been studied in many regions in Monsoon Asia.

Reviewing these studies for all of Monsoon Asia, an attempt has been made to clarify the distribution of monthly precipitation in summer in connection with monsoon circulation from the standpoint of synoptic climatology in another paper in this monograph by the present writer and his co-worker.

For further investigations, precipitation data must be obtained and analyzed from region to region. The studies on the seasonality of rainfall by Fitzpatrick et al. (1966) and of detailed rainfall distribution by Brookfield and Hart (1966) both for the tropical Southwest Pacific, are good examples. These reveal detailed rainfall conditions as well as variability, intensity and spells of weather with sufficient data for more than 600 stations in New Guinea and its surrounding districts. Little is known at present on the water balance in the mountain regions in Monsoon Asia. Efforts to collect basic data on precipitation and runoff and evaporation, such as have been made in the Hindu Kush by Flohn (1969), will probably extended also to other regions.

Ivanov (1958) has published a comprehensive monograph on atmospheric moisture in tropical and adjacent countries. His discussion of the annual precipitation pattern and types and seasonal variation of precipitation as well as his climatic classification according to the annual pattern of atmospheric moisture balance are quite valuable. A well selected bibliography of rainfall in the Indo-Pakistan subcontinent was recently published (Domrös, 1970).

3. Water Balance Expressed by Climatic Indices

Climatic indices are an important means of expressing the climatic state in classic climatology. Comprehensive reviews of such indices have been presented by Knoch and Schulze (1954), Blüthgen (1966), and Burgos (1968). Of the various indices expressing degrees of wetness and dryness, the following have been found most useful for many regions.

a)   The rain factor (RF) is expressed by Lang (1915) as:

$$(RF) = \frac{P}{T_{15}}, \tag{1}$$

where P = annual precipitation (mm) and T_{15} = average of mean temperature during frostless period. Because of the difficulty in obtaining the T_{15} value for each station, the following revised formula was presented:

$$(RF) = \frac{P}{T_{20}}, \tag{2}$$

where T_{20} = (1/12) (summation of the monthly mean temperature above 0°C).

b)   The aridity index (AI) is expressed by Martonne (1926) as:

$$(AI) = \frac{P}{T+10}. \tag{3}$$

Here T stands for the annual mean temperature, and this formula has been applied in many regions of the world. The boundary line of arid conditions is given by Köppen (1936) as:

$$\frac{P}{T+\alpha} = 20. \tag{4}$$

Humid-arid boundary using the annual evaporation, annual precipitation and annual mean temperature was discussed for East Asia (Fukui, 1943). The secular change of climate expressed by Köppen's method was studied in the areas surrounding the North Pacific Ocean (Fukui, 1965). Detailed maps using the original method of the Köppen's classification and revised maps using the year climate method are presented by Mizukoshi in this volume. A revised formula for (1) and (2) is given by Ångström (1936):

$$\frac{P}{1.07^T}. \tag{5}$$

These indices, expressed by air temperature and precipitation (excluding Köppen's), are applied for Monsoon Asia and discussed later in this volume by Kawamura, because the distributions of these indices have been studied mainly in regions in the temperate zone and as yet only little in the tropical or subtropical zones.

c)   The indices connecting precipitation and evaporation or saturation deficit were given by Meyer (1926) as:

$$\text{Annual } (P-S) \text{ ratio} = \frac{P}{S}, \tag{6}$$

where P is the annual precipitation (mm) and S is the annual saturation deficit (mb). A modified form by Prescott (1958) is

$$P/S^{0.75} \quad \text{or} \quad P/E^{0.75}, \tag{7}$$

where E is the annual evaporation amount (mm). The usefullness of these indices varies according to the objectives: Namely, to express the natural landscape, soil and vegetation distribution, agricultural land use and so on.

d) Thornthwaite (1931, 1933) devised the famous PE index (I):

$$I= \sum_{n=1}^{12} \left(\frac{p}{e} \right), \tag{8}$$

where p is the monthly precipitation (mm) and e the monthly evaporation (mm). Instead of e, formula (8) can be expressed by

$$I= \sum_{n=1}^{12} 0.12 \left(\frac{p}{t+12.2} \right)^{\frac{10}{9}}, \tag{9}$$

where t is the monthly temperature (°C). If the PE index is more than 128, the vegetation is rain forest; at 64–127 it is forest, at 32–63 grassland, at 16–31 steppe and at 0–15 desert. This PE index (precipitation effectiveness) is combined with temperature efficiency types and with four subtypes of seasonal distribution of effective precipitation based on Thornthwaite's old classification system. Stamp's opinion (1962) on classification in Asia according to this system is as follows: "It is by no means certain that it will be acceptable to the student of the geography of Asia, because it produces too many strange bed-fellows and there are too many anomalies in Asia."

e) Thornthwaite (1948) developed a concept of potential evapotranspiration and attempted to describe climate by comparing its amount with the precipitation. The results of this theory as applied to Monsoon Asia are discussed by Kayane in this volume. He presents, among other things, a map of hydrological regions in Asia combining water balance components such as the annual water deficit and surplus obtained by Thornthwaite's new method.

4. WATER BALANCE FROM THE SYNOPTIC CLIMATOLOGICAL VIEWPOINT

Much can be learned concerning water balance in Monsoon Asia from the standpoint of synoptic meteorology and climatology from the book by Trewartha (1962). Before the publication of this book, there were many milestones in the development, such as the work of Flohn (1950), Thompson (1951), Schick (1953), Watts (1955), Flohn (1957), Staff Members of the Academia Sinica (1957, 1958) and Kurashima (1959). Since the publication of Trewartha's climatography, striking progress has been made. Only work done in the past ten years is reviewed here.

Chinese contributions of this period are treated first. A monograph edited by Kao (1962) contains seven papers on problems concerning the monsoons in East Asia. Among these, Kao and Hsü (1962) dealt with the duration of the rainy season in relation to the monsoon in East Asia. Table I was compiled from the results for China and adjacent regions by them, for India by Chatterjee (1953) and for Japan by Yoshino (1965a). Of interest is that the rainy season begins earliest in South China, at the beginning of May. Then the date line gradually extends north to the Pacific coast of southeastern Japan. On the other hand, the arrival of the rainy season is delayed in South Asia, and comes only in early or mid-June in

Table I. Beginning of the rainy season in Monsoon Asia (Compiled from the results after Chatterjee, 1953; Kao and Hsü, 1962; Yoshino, 1963, 1965a, b).

	75	80	85	90	95	100	105	110	115	120	125	130	135	140 °E
45°N	—	—	—	—	—	—	—	—	—	Aug. 9–13	July 15–19	July 15–19	—	—
40	—	—	—	—	—	—	—	July 20–24	July 20–24	July 20–24	July 15–19	—	—	June 25–29
35	—	—	—	—	—	—	July 20–24	July 20–24	July 15–19	July 10–14	July 10–14	June 20–24	June 15–19	June 15–19
30	June 30 –July 4	June 15–19	June 10–14	May 31 –June 4	—	June 5–9	June 25–29	June 20–24	June 15–19	June 15–19	June 15–19	June 5–9	May 26–30	May 21–25
25	June 15–19	June 10–14	June 10–14	June 5–9	May 31 –June 4	May 31 –June 4	May 31 –June 4	June 10–14	June 5–9	May 30 –June 4	—	May 15–20	—	—
22.5	June 10–14	June 10–14	June 10–14	June 5–9	May 31 –June 4	May 21–25	May 21–25	May 16–20	May 6–10	May 6–10	—	—	—	—
20	June 10–14	June 10–14	June 5–9	May 26–30	May 26–30	May 21–25	May 1–5	May 1–5	—	—	—	—	—	—

southern India. According to Kao and Hsü (1962), the duration of the rainy season on the average is from May 19 to September 2 in South China, June 20 to August 24 in Middle China, and July 20 to August 19 in North China. These have, of course, a close relation to the summer monsoon, whose duration on the average is from April 26 to September 27 in South China, June 10 to September 12 in Middle China, and July 10 to September 2 in North China. In other words, the rainy season begins about ten days later than the summer monsoon and ends two or three weeks earlier.

Dao *et al.* (1965a) published a monograph on subtropical weather systems in summer in China: typhoons, heavy rainfall during the rainy season and subtropical anticyclones in the Northwest Pacific. In particular, water vapor transport, one of the basic conditions of heavy rainfall, was considered in relation to the southerly component of air flows at the 700 or 850 mb levels and the regional coincidence was pointed out (Dao *et al.*, 1965b).

Tiri-chi-kan No. 9, published in Peking in 1965, was devoted to a collection of studies from the standpoint of synoptic climatology: studies on rainfall characteristics in the rainy season,

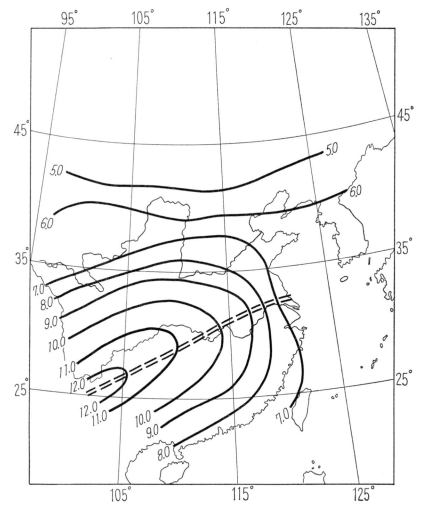

Fig. 1. Mean specific humidity (g/kg) at the 700 mb level in July, 1954 (after Shen, 1965).

the summer monsoon in China in relation to the circulation systems in both hemispheres, the winter monsoon and its related weather and circulation and so on. In this monograph, Hsü and Hsu (1965) presented composite maps of topography at the 500 mb level and the pressure at sea level during abnormally wet and dry years in China. It was shown that when the center of the polar high in spring is located in the sector between 140°E and 160°W in the polar region north of 65°N, the rainfall during the rainy season in China shows a plus anomaly; when the center is located in the sector between 140°W and 90°W, a minus anomaly is seen. This result corresponds to the observation obtained by Wada (1962) that when the polar vortex deviates to the Western Hemisphere in spring, the cold air tends to flow into the North American continent and brings drier conditions in China in the rainy season and, conversely, when the polar vortex deviates to the Eastern Hemisphere, wetter conditions occur in the rainy season over East Asia. Shen (1965) analyzed the three dimensional structure of the dry and wet conditions during the rainy season. Figure 1 shows the distribution of mean specific humidity at the 700 mb level in July, 1954, when the rainy season was remarkably strong. The double broken line shows the center line of the so-called moist tongue. As a reference, the distribution of precipitation anomaly (mm) for the same month given by Chen (1957) is reproduced in Fig. 2. The region with a mean specific humidity greater than 9 g/kg corresponds roughly to the region with a plus anomaly of precipitation. On the other hand, when the precipitation showed a minus anomaly, -50%, as in July, 1959, the water vapor transport calculated from sea level to the 300 mb level was -30 to -50% of the normal year conditions in China. The water vapor transport will be discussed latter in detail.

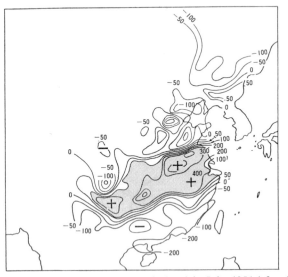

Fig. 2. Distribution of precipitation anomaly (mm) in July, 1954 (after Chen, 1957).

The circulation characteristics during the periods of persistent drought and flood in the Yangtze and Hwai-Ho valleys in July 1954, 1959, and 1961, have also been analyzed (Dao and Hsu, 1962). Using the precipitation data for 60 years, 1901–1960, eleven types of precipitation anomaly distribution in the broad region of eastern China were distinguished (Chang et al., 1963). It was shown that during recent years the negative anomaly has been predominant.

The precipitation distribution in August was considered in relation to geostrophic wind conditions by Kuo and Yang (1965). The relationship was also studied for four stages of the rainy season over East Asia. The following are results published in earlier papers by the present writer (Yoshino, 1963, 1965b, 1966).

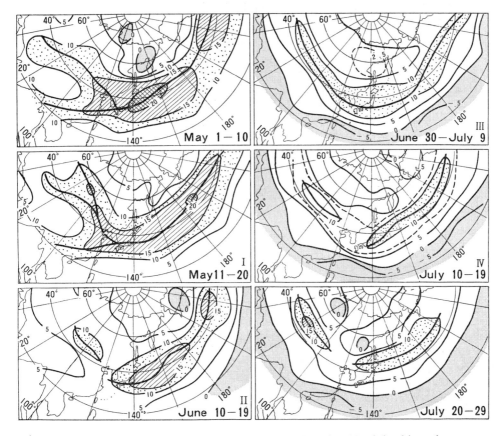

Fig. 3.   Distribution of geostrophic wind velocity (m/s) at the 500 mb level in each stage over East Asia. Westerly component: positive (after Yoshino, 1965b).

The rainy season in early summer over East Asia, called the Mai-yü in China, Mae-ue in the southern part of Korea and Baiu in Japan, was divided into four stages by considering the characteristics of the westerly zonal winds at the 500 mb level (Fig. 3) and the positions of the frontal zones at sea level (Fig. 4). The rainfall distribution in each stage is shown in Fig. 5. *Stage I:* An apparent frontal zone, though its west-east length is no longer than several hundred kilometers, appears in southern China or along the Pacific coast of Japan. The axis of the subtropical westerly wind maximum runs at about 35°N. This stage begins mostly towards the middle or end of May. A belt-like rainfall region from Taiwan to the northeast is observed. *Stage II:* The frontal zone begins to develop from central China to the Pacific coast of Japan. At this stage the axis of the polar westerly maximum comes down south and joins the axis of the subtropical westerly maximum. The rainy season begins from the Yangtze River region to southwestern Japan in most years in the middle of June. The rainfall zone, staying at the same position as in Stage I, increases the rainfall. The maximum

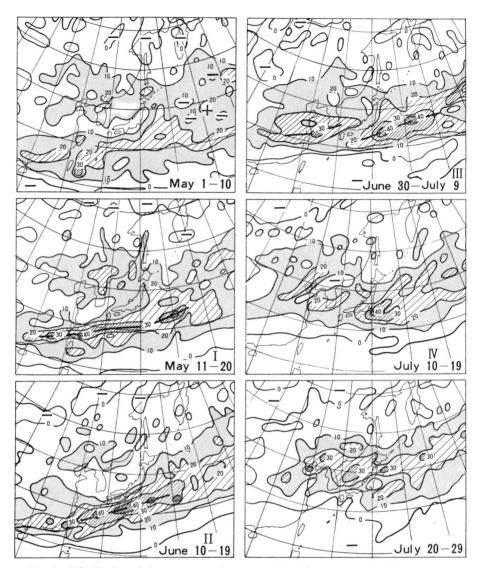

Fig. 4. Distribution of the occurrence frequency (%) of fronts at sea level in each stage (after Yoshino, 1965b).

concentration area shows 250 mm/10 days. *Stage III:* A marked frontal zone with a frequency of occurrence of 40–70% develops from central China to Japan over a distance of 4,000–5,000 km. The axis of the westerly wind maximum runs along 35°N with blocking high in the Sea of Okhotsuk. This stage is the climax of the rainy season. The marked rainfall zone develops from central China to Kyūshū in southwestern Japan. *Stage IV:* The frontal zone at sea level sometimes remains in China, but is obscure and shifts to the north over East Asia as a whole. The axis of the westerly wind maximum retreats to 45°N or further at the 500 mb level. As far as the zonal wind pattern at the 500 mb level is concerned, the rainy season is over by this stage. There are two rainfall zones, the first situated along the Pacific coast of western Japan, continuing from the preceding stage, and the second located

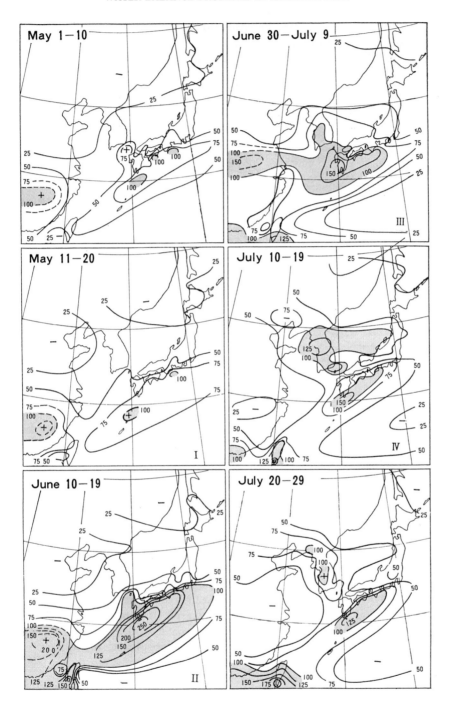

Fig. 5. Distribution of precipitation (mm) in each stage (after Yoshino, 1966).

from the middle part of the Korean Peninsula to the Tōhoku district along the Japan Sea coast. Japan often suffers from grave disasters caused by local heavy rainfall in this stage. For instance, an extraordinary record, 1,109.2 mm, was observed for 24 hours on July 25–26, 1957, at Saigo, Nagasaki Pref., Japan.

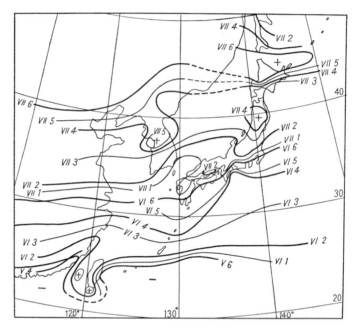

Fig. 6. The appearance of the rainfall maximum. V, VI and VII equal May, June and July. 1, 2, 3, ..., 6 equal the first, the second, the third, ..., the sixth half-decade of each month (after Yoshino, 1966).

The appearance of maximum rainfall during the rainy season is presented in Fig. 6. The pattern shifts clearly from south to north as was demonstrated in Table I, which gave the shift of the rainy season over Asia. This fact does not contradict the fact of the stagnancy of the rainfall zone along the Pacific coast of southwestern Japan shown in Fig. 5. This situation can be explained in that the maximum rainfall during the rainy season at the northern stations is smaller than the same time values at the stations located at 23–25°N on 120°E, at 30°N on 130°E and at 34°N on 140°E, if we compare their values along the respective longitudes. It can therefore be seen that the rainfall zone stays almost at the same location throughout the stages in the rainy season.

The monsoon circulations in South and East Asia have been reviewed in detail by Kurashima (1968) and Kurashima and Hiranuma in another section of this volume.

Some new facts concerning the subtropical anticyclone over the Northwest Pacific in the lower troposphere and the anticyclone over the Tibetan Plateau in the higher troposphere are also discussed by Wada in this volume. The relation between summer precipitation in Monsoon Asia and large-scale circulation patterns at the 500 mb level is studies from the correlation field and composite maps. Neyama deals with the seasonal change of the saturation rate distribution in connection with the circulation pattern at the 850 mb level.

5. Water Vapor Transport in Monsoon Asia

The basic studies on water vapor transport in Monsoon Asia were made by Flohn and Oeckel (1956) and Murakami (1959). The latter calculated the water vapor transport in ten-day totals during the rainy season in 1957 at 125 points over East Asia. He found that water vapor in the early stages is transported mainly from the southwest by an airflow coming from Southeast Asia with the concentration at the 700 mb level, but in the last stage the moisture flow, originating from the Pacific and taking a roundabout route over South China and the East China Sea, becomes more significant in the lowest layer below the 850 mb level.

The water vapor transport and water balance over eastern China were calculated using data from January and July 1956 by Hsü (1958). It was found that, even though the transport of water vapor in summer was quite different from that in winter, the inflow of water vapor from the south was much more important. It was also pointed out that the continent was a source region of water vapor and that evapotranspiration was larger than precipitation, at least in July, 1956.

Table II. Water vapor transport in the upper-Yangtze region, average for 1955–1961. Unit: km³ (water equivalent) (after Wang and Xu, 1964).

Boundary	Layer							Inflow	Outflow	Net transport
	Surface –900	900– 850	850– 800	800– 700	700– 600	600– 500	500–400 mb			
East	2.6	8.1	16.3	43.4	35.8	19.7	9.6	4.5	140.0	−135.5
South	10.5	36.5	54.2	110.5	72.3	32.0	6.2	323.6	1.4	322.2
West	0.0	0.0	0.5	5.6	6.0	8.3	14.8	36.9	1.7	35.2
North	0.0	2.8	9.2	40.4	31.8	13.4	0.5	4.4	102.5	−98.1
Total								369.4	245.6	123.8

The water balance in the basin (500,000 km²) of the upper Yangtze river in summer was investigated for 1955–1961 by Wang and Xu (1964). An average of six years, 1955–1961, is given in Table II. The inflow of water vapor from the south is 323.6 km³, or 83% of the total inflow, against the small amount of inflow from east and north. The inflow from the south does not change significantly, but the outflow over the east and north boundaries change quite differently from year to year. The net transport is 123.8 km³ and the runoff is 109.3 km³. Maximum water vapor transport was found in the 700–800 mb layer. For average conditions, the components of the hydrological cycle, according to the system by Budyko and Drozdov (1953), were obtained as follows (Fig. 7): If we take the total inflow (F_i), 369.5 km³, as 100% it can be estimated that precipitation from water vapor originating advection (r_a), 45.3%, and precipitation from water vapor originating within the region (r_i), 4.3%, makes a total precipitation (R), of 49.5%. The evaporation (E) is estimated as 18.5%. The difference, $E-r_i$, is 14.2%. This 14.2% plus (F_i-r_a), 54.7%, makes a total outflow of water vapor F_0, of 68.9%. On the other hand, the surface runoff (F_s) is 29.6%. Therefore, $100.0-68.9-29.6=1.4\%$ can be given as the gound water runoff. The coefficient of water cycle $K=R/r_a$ is calculated as 1.094 in this region and this is larger as compared

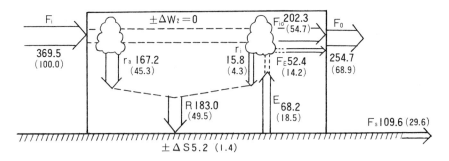

Fig. 7. Hydrologic cycle in the upper Yangtze region (500,000 km²). Unit: km³. Values in brackets are percent, making $F_i = 100.0\%$ (after Wang and Xu, 1964).

with the values, $K = 1.075$, in the whole Yangtze region in summer and also for a year, or 1.079 in the Mississippi River region for a year, obtained previously by Cheng and Shen (1959).

Saito (1966) studied the water balance associated with the monsoon wind systems in the rainy season in East Asia. Concerning the net flux it was shown that the contribution of the northward net flux becomes of the greatest importance as the season advances in Japan, even though the eastward flux on the western boundary has the largest magnitude as was indicated by Flohn and Oeckel (1956). From the results, it was also suggested that a considerable amount of vapor is necessary on the continental part of East Asia.

Sorochan has made very detailed analyses of the hydrologic cycle under the influence of monsoon circulation in eastern Siberia and East Asia (Drozdov and Grigor'eva, 1965). One of the important results, transport velocity, $\bar{u}$ (m/s), taking into account trajectory curvature, moisture content up to 7 km, w (kg/m²), and the effective transport height, H (km), is given in Table III. From this table it can be seen that the annual mean moisture content in the zone 40–50°N over East Asia is 14 kg/m²; which is slightly smaller than that over the European USSR, 15 kg/m²; but the annual range, i.e. the concentration in the months during the summer monsoon, 35–36 kg/m², is much greater over East Asia, than the 27 kg/m² over the European USSR.

The meridional and the zonal components of the mean total water vapor transport field for the Northern Hemisphere (Starr and Peixoto, 1958) and its eddy flux (Starr and Peixoto, 1964) have been calculated.

These rescarchers made clear, for instance, that the meridional transport of water vapor by the Hadley circulation in the tropics does not imply a comparable importance of this cell for hemispheric momentum and zonal kinetic energy considerations. The distribution map showing the time average of the vertically integrated specific humidity, i.e. precipitable water, calculated for 1958 over the whole Northern Hemisphere revealed that the region including the Arabian Sea, southern India and the Bay of Bengal has a value of 5 g/cm², which is the highest in the world. In contrast, the value is 0.5 g/cm² over the Tibetan Plateau, the smallest value as compared with those on the same latitude (Starr *et al.*, 1965). The vector field of the vertically integrated water vapor transport for 1958 shows that the largest values, $21–24 \times 10^2$ g/cm/sec, are found at the southwestern parts of the subtropical anticyclones over the North Pacific and the North Atlantic. The large values, directed eastwards, are found over Japan, due to the strong subtropical westerly jet stream. Over the Tibetan Plateau, the southerly component is predominant.

Table III. Annual variation of hydrologic-cycle components for eastern Siberia and the Far East, according to Sorochan (after Drozdov *et al.*, 1965).

Zone (°N)	Hydrologic-cycle components	Jan.	Feb.	Mar.	Apr.	May	June	July	Aug.	Sept.	Oct.	Nov.	Dec.	Year
1 60–70°	$\bar{u}$ (m/sec)	—	9.0	8.1	7.0	6.2	6.5	6.0	6.1	6.4	6.4	8.1	8.1	7.9
	w (kg/m²)	—	3.6	4.4	6.0	9.3	16.2	22.4	20.1	13.6	8.0	4.6	3.7	9.8
	H (km)	—	3.4	3.3	2.1	1.7	2.2	1.6	1.3	1.5	1.5	2.1	2.6	2.4
2 50–60°	$\bar{u}$ (m/sec)	9.2	8.7	7.6	7.9	7.7	6.7	6.1	6.6	7.0	8.2	8.9	8.3	7.7
	w (kg/m²)	3.3	3.4	5.7	8.0	12.2	23.5	31.8	29.6	17.8	9.7	4.9	4.0	12.4
	H (km)	2.4	2.4	1.5	1.6	2.0	2.0	1.8	1.8	1.5	1.5	1.7	1.8	1.8
3 40–50°	$\bar{u}$ (m/sec)	11.5	10.3	10.6	9.8	9.9	7.7	7.1	7.6	8.4	9.8	11.8	12.5	8.9
	w (kg/m²)	3.2	3.3	5.1	9.9	15.4	25.6	36.0	35.1	21.8	12.0	6.3	3.7	14.2
	H (km)	2.4	2.4	2.2	1.9	1.4	1.7	2.0	1.8	1.6	1.6	2.0	2.5	1.9

A very detailed map of horizontal divergence of the vertically integrated water vapor transport revealed that heavy divergence occurs in the Bay of Bengal and the Indian Ocean, indicating a strong excess of evaporation as compared with precipitation. Such maps, which give a detailed regional condition of $(E - P)$, can be made only from aerological data, because the data observed at meteorological stations are very limited in the oceans.

This method of calculating the divergence fields can also be applied to study the water balance in Monsoon Asia. In this volume, Asakura presents the results obtained from data from October, 1968, to September, 1969. A striking condition in Monsoon Asia is that the precipitable water makes up about 20% of the total amount in the Northern Hemisphere, reaching a maximum 28% in July. The hemispheric distribution of water vapor divergence is presented by Kubota in this volume. The seasonal change of the sink and source regions in Monsoon Asia will be discussed in detail.

6. CLIMATIC FLUCTUATION

a. *Geological Time*

Few studies have been made on climatic change in geological times in Monsoon Asia. A first tentative summary in Asian palaeoclimatology was published by Kobayashi and Shikama (1961). The climate of the early Palaeozoic is thought to have been warm or even hot. In the late Palaeozoic, Jido (the lower middle period in Korea) climate was warm and humid, while Kobosan (the upper middle period in Korea) sediments were considered to be deposited in a fairly arid, warm, non-seasonal climate. The Mesozoic climate in Japan became first hotter and then somewhat cooler towards the later stages.

At the beginning of the Cenozoic, a rather dry climate is suggested by the red and reddish sandstones in the upper Yangtze, Kwangtung and Kwangsi. The change from reddish to yellow sediments is thought to have been occasioned by a climate with strong prevailing winds. South China and Japan were covered by loamy materials, but the Tsingling range is situated on the border between the loessic and loamy regions. Wind effects on the pebbles are dominant in Japan, suggesting strong winds in a periglacial region. The climatic change curve in the Cenozoic in East Asia shows that from the early Eocene to the early Miocene, warm temperatures decreased to temperate and then, from the early to middle Miocene, became again subtropical, hot and humid. After that a rainy climate, foggy in summer and mild in winter, is surmised in Japan. It is known that the climate in the Pleistocene fluctuated strikingly, ranging between humid-temperate and arid-cold conditions. For instance, climate in the latest Pleistocene or the earliest Holocene was semiarid and 6–8°C colder than at present. Then it became humid and warmer, but the warming tendency is thought to be decreasing. The climatic change in the Quaternary will be described in detail by Tabuchi and Urushibara later in this volume. In their review paper, it will be shown that a climatic optimum around 6,000 B.P. and a climatic deterioration around 2,500 B.P. occurred simultaneously in Monsoon Asia. But it was arid in both periods in Central Asia, humid in both periods in the Near East, and humid in the former period and arid in the latter in West Pakistan.

An attempt to reconstruct the atmospheric circulation patterns in the Quaternary age was made by Lamb (1961). According to his tentative maps of the mean surface pressure distribution over the Northern Hemisphere, the following can be interpreted as conditions in the ice age in Monsoon Asia. In winter, the low in the North Pacific showed 995 mb, which is 5 mb lower than at present, with the center located slightly west as compared to the present time. In contrast, the high in Siberia developed rather weakly, with its center at 1,025 mb, 10 mb lower than at present. Therefore, the northwesterly winter monsoon in East Asia and the northeasterly in Southeast Asia are thought to have been weaker in the ice age than at present. As Central Asia was arid in the ice age, dry conditions could be expected over East Asia also, even though the monsoon winds were weaker. The Pacific polar frontal zone, however, might have developed more strongly in the ice age in winter, because cyclonic activity in the north Pacific was stronger, as indicated by a lower barometric pressure in the North Pacific.

The difference between the circulation pattern at present and the supposed circulation pattern in the maximum Quaternary glaciation period is more apparent in summer than in winter. In the ice age, the low pressure area in South Asia had two centers with 995mb in summer, one located in eastern India and another in the Persian Gulf, against present 1,000 mb. This implies a stronger SW monsoon over South and Southeast Asia in the ice age, and more rainfall. At 50°N on 120°E, 50°E, and 170°W, there were three low centers with 1,000 mb or 1,005 mb in the age, which means a stronger frontal activity at this latitude, but Central Asia had a relatively higher pressure. The North Pacific anticyclone expanded slightly westward and accordingly, the southerly or southwesterly summer monsoon was markedly developed over East Asia in the ice age. The NITC over the Northwest Pacific was probably located slightly south near the Philippines and over the South China Sea. The NITC over India should also have been apparent in the ice age.

b. *In Recent Years*

The data on rainfall since 1770 in Seoul has been analyzed by Tada (1938) and Arakawa (1956). The former found, through harmonic analysis, cycles with 2, 6, 13, 35, and 37 years in the annual rainfall. Yamamoto (1952) showed clearly that the summer rainfall curve at Seoul had a fairly good correspondence to the curve of the number of fine days in Tokyo from the 18th century up to the present. Moreover, the curve of a ten-year running mean of the total amount of rainfall in June, July and August in Nagano, Central Japan, shows a surprisingly good negative correlation to that of the July rainfall at Lahore, West Pakistan, and of the rainy days in summer in Tokyo (Yamamoto, 1958). The regional correlation of the secular change of rainfall over East Asia was clarified by calculating the correlation index (Yoshino, 1963). For the whole region in Monsoon Asia, Yoshimura made a regional division according to the tendency of secular change, applying the same method; the results are given elsewhere in this volume. The amount of rainfall in July in the region along the Pacific polar frontal zone seemed greater when the meridional component of the general circulation of the atmosphere in the Northern Hemisphere prevailed and was less significant when the zonal component dominated.

From rainfall data over a period of eighty years (1885–1963), the secular change in the Mai-yü was studied for the middle and lower Yangtze region (Hsu, 1965). According to Hsu's table, the earliest onset of Mai-Yü occurred on May 26, 1896, which is 19 days earlier than normal, the latest onset occurred on July 4, 1947. The earliest and the latest terminal dates were June 16, 1961, which is 23 days earlier than normal, and August 1, 1954. The longest Mai-Yü was in 1896, and lasted for 87 days with +183% of the normal rainfall. In 1954, the Mai-yü was also abnormal: it lasted 50 days longer than normal with a +202% rainfall. Exceptionally dry years were 1893, 1897, 1898, 1902, 1904, 1925, 1934 and 1958. The secular change of data for the set-in of the Mai-yü shows a clear 22-year cycle as given in Fig. 8.

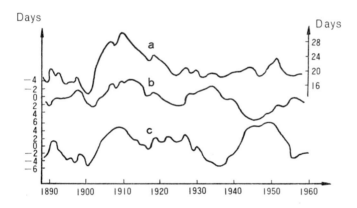

Fig. 8. Secular change of the Mai-yü in the middle and lower Yangtze regions, 10-year running mean (after Hsu, 1965). a: Duration deviation, b: Date of onset, c: Date of end. Minus sign means earlier.

Floods and droughts in India have been studied intensively and were the theme of a symposium taking place in 1958 (Basu, 1959). Many valuable contributions were presented in the proceedings, including an interesting paper on the secular change of length of the wet season in Bihar and the eastern districts of Uttar Pradesh (Mallik, 1958). There appeared to be no periodicity or general tendency in the recurrence of bad years. There found no bad years occurred at the same time as in East Asia, except in 1925.

Ten Hoopen and Schmidt (1951) reported a high negative correlation in precipitation change at Pontianak, situated on the equator, and in West Java. This interdependence was thought to be caused by the outcome of the southward displacement of the ITC in autumn and winter in the Northern Hemisphere. Such interactions between the middle latitude and the tropics should be studied in detail in the future, particularly for Monsoon Asia.

7. WATER BALANCE AS RELATED TO WATER RESOURCES

In this volume, specific studies on water resources in Monsoon Asia are not presented. We will, therefore, only review here some problems relating to recent results of certain studies.

Ienaga (1969) has dealt with the problems of water resources in Monsoon Asia with special emphasis on agricultural development: In Monsoon Asia, water in paddy fields is obtained in most cases directly from rain-water, because there are no irrigation systems. Cultivation of rice therefore depends primarily upon the rainfall in the monsoon season. In terms of temperature conditions alone, rice could be grown throughout the year but because of the dependency on rainfall, it is cropped only once a year in most areas. Two successive rice crops are obtained where irrigation systems have been sufficiently developed or enough rainfall is obtained from both the southwest and northeast monsoons. Due to these circumstances, rainfall in the monsoon seasons should be studied in detail from region to region, especially its fluctuations from the standpoint of climatology.

The Center for Southeast Asian Studies at Kyoto University held a symposium on agricultural techniques in 1967. At this symposium, problems of irrigation systems and their relation to rainfall in Thailand (Sawada, 1968), Malaysia (Fujioka, 1968) and in the Philippines (Nishiguchi, 1968) were brought up. Further studies are needed on the rainfall pattern, in particular on the basis of climatic water balance problems and also from the synoptic climatological viewpoint.

One of the more rational approaches to the climatological representation of water balance is Thornthwaite's method, mentioned earlier. In a study on the climatic division of Thailand, this method was applied (Ogino, 1967) to make clear the distribution of humid and arid conditions. Subrahmanyam (1958) also dealt with droughts and aridity in India by Thornthwaite's method taking into consideration evapotranspiration and precipitation. He concluded that it was necessary, in the day-to-day application of irrigation, to work out the moisture balance on a daily basis. The water balance method is a very powerful tool for irrigation scheduling.

Finally, the problem of satellite cloud photographs must be considered. These photographs have been used in recent years to determine the mean surface position of the ITC and also the region of water vapor transport as demonstrated in the study by Asakura in this volume. As has been pointed out by Sadler (1969), it will be important to find higher correlations in the distributions between average cloud cover revealed by satellites and mean rainfall. When this can be achieved a detailed distribution map of rainfall over the oceans could be made and water resources over the oceanic regions, which at present are little known, could be clarified.

References

Albrecht, F. 1951: Monatskarten des Niederschlags im Indischen und Stillen Ozean. *B. D. W. US-Zone* **4**(29) 21S.

Anathakrishnan, R., Rajagopalachari, R. J. 1964: Pattern of monsoon rainfall distribution over India and neighbourhood. *In* Proc. Symp. Trop. Met. 1963, 192–200.

Ångström, A. 1936: A coefficient of humidity of general applicability. *Geogr. Ann.* **18** 245–254.

Arakawa, H. 1942: Daitōa no Kikō (Climate of East Asia). Asahi Shimbun Co., Tokyo, 199p.

Arakawa, H. 1956: On the secular variation of annual totals of rainfall at Seoul from 1770 to 1944. *Arch. Met. Geoph. Biokl.* (*B*) **7**(2) 205–211.

Arakawa, H. (ed.) 1969: Climates of northern and eastern Asia. *World Survey of Climatology* **8** 1–248.

Basu, S. 1959: Proceeding of the Symposium on Meteorological and Hydrological Aspects of Floods and Droughts in India. India. Met. Dept. 205p.

Blüthgen, J. 1966: Allgemeine Klimageographie. Walter de Gruyter, Berlin, 720S.

Brookfield, H. C., Hart, D. 1966: Rainfall in the tropical Southwest Pacific. Res. Sch. Pacific Stu., Dept. of Geogr. Publ. G/3, Austr. Nat. Univ., Camberra 25p+Maps+Figs.

Budyko, M. I., Drozdov, O. A. 1953: Zakonomernosti vlagooborota v atmosfere (Regularities of the hydrologic cycle in the atmosphere). *Doklady AN SSSR* **90**(2) 167–170.

Burgos, J. J. 1968: World trends in agroclimatic surveys. *In* Agroclimatological Methods, Proc. Reading Symp. Agroclimatological Methods (UNESCO) 211–224.

Chang, Hs.-k., Kong, Y., Hsu, Ch. 1963: A preliminary analysis of abnormal precipitation distribution in eastern China in the first six decades of the present century. *Acta Met. Sinica* **33**(1) 64–77.

Chatterjee, S. B. 1953: Indian Climatology. Calcutta, 417p.

Chen, H.-ë. 1957: Characteristics of atmospheric circulation and its relation to the flood on the Yangtze River basin and Huaihe in the Chinese People's Republic in 1954. *Acta Met. Sinica* **28**(1) 1–12.

Cheng, S.-ch., Shen, Ch.-ch. 1959: Atmospheric water cycle in the Yangtze region. *Acta Geogr. Sinica* **25**(5) 346–355.

Daigo, Y. 1947: Handy Atlas of Agricultural Meteorology in Japan. Kyoritsu Pub. Co., Tokyo, 234p.

Dao, Sh.-y., Hsu, Sh.-y. 1962: Some aspects of the circulation during the periods of the persistent drought and flood in Yangtze and Hwai-Ho valleys in summer. *Acta Met. Sinica* **32**(1) 1–10.

Dao, Sh.-y. *et al.* 1965a: Studies on Some Problems on the Subtropical Weather Systems in Summer in China. Peking, 146p.

Dao, Sh.-y. *et al.* 1965b: Study on the short range forecasting of heavy rainfall in the region on the middle and upper course of the Yangtze valley. *In* Some Problems on the Subtropical Weather Systems in Summer in China. 59–77.

Domrös, M. 1968. Zur Frage der Niederschlagshäufigkeit auf dem Indisch-Pakistanischen Subkontinent nach Jahresabschnitten. *Met. Rdsch.* **21**(2) 35–43.

Domrös, M. 1970: A rainfall atlas of the Indo-Pakistan Subcontinent based on rainy days. *In* Studies in the Climatology of South Asia (Schweinfurth *et al.*) 7+Figs.

Domrös, M. 1970: Bibliography (1945–1969) of rainfall conditions in the Indo-Pakistan Subcontinent. *In* Studies in the Climatology of South Asia (Schweinfurth *et al.*) 9–16.

Drozdov, O. A., Grigor'eva, A. S. 1965: The hydrologic cycle in the atmosphere. Israel Progr. f. Sci. Transl., Jerusalem, 282p.

Fitzpatrick, E. A. *et al.* 1966: Rainfall seasonality in the tropical southwest Pacific. *Erdkunde* **20** 181–194.

Flohn, H. 1950: Studien zur allgemeinen Zirkulation der Atmosphäre. *Ber. D. Wetterd. U.S.-Zone* (18) 1–52.

Flohn, H., Oeckel, H. 1956: Water vapour flux during the summer rains over Japan and Korea. *Geoph. Mag.* **27** 527–532.

Flohn, H. 1957: Large-scale aspects of the summer monsoon in South and East Asia. *75th Ann. Vol. J. Met. Soc. Japan* 180–186.

Flohn, H. 1969: Zum Klima und Wasserhaushalt des Hindukuschs und der benachbarten Hochgebirge, *Erdkunde* **23** 205–215.

Flores, J. F., Balagot, V. F. 1969: Climate of the Philippines. *World Survey of Climatology* **8** 159–213.

Fujioka, Y. 1968: Agricultural development by means of irrigation and drainage in Southeast Asian countries. *The Southeast Asian Studies* **5**(4) 793–803.

Fukui, E. 1942: Nanpôken no Kikô (Climate of Southeast Asia). Tokyo-do Book Co., Tokyo, 316p.

Fukui, E. 1943: "Trockengrenze" in Eastern Asia. *J. Met. Soc. Japan* **21** (12) 537–541.

Fukui, E. 1965: Secular shifting movements of the major climatic areas surrounding the North Pacific Ocean. *Geog. Rev. Japan* **38** (5) 323–342.

Hann, J. 1910: Handbuch der Klimatologie. Band II: Klimatographie, Teil I: Klima der Tropenzone, Stuttgart, 426S.

Ten Hoopen, K. J., Schmidt, F. H. 1951: Recent climatic change in Indonesia. *Nature* **168** 428–429.

Hsu, Chun 1965: An analysis of Mei-yü in the middle and lower Yangtze valley of recent eighty years. *Acta Met. Sinica* **35**(4) 507–518.

Hsü, Sh.-y. 1958: Water-vapour transfer and water balance over Eastern China. *Acta Met. Sinica* **(29)** 33–43.

Hsü, Sh.-y., Hsu, M.-y. 1965: Atmospheric action center and the prolonged wetness and dryness in the Kiang Hwai region in summer. *Tiri-chi-kan* (9) 1–18.

Ienaga, Y. 1969: Tônan Ajia no Mizu (Water of Southeast Asia). Ajia Keizai Kenkyu-jo, Tokyo, 161p.

Ivanov, N. N. 1958: Atmosfernoe uvlazhnenie tropicheskikh i sopredel'nykh stran zemnogo shara (Atmospheric Moisture in Tropical and Adjacent Countries). Akad. Nauk SSSR, Leningrad, 311p.

Johnson, B. L. C. 1969: South Asia, Selective Studies of the Essential Geography of India, Pakistan and Ceylon. Heinemann Educational Books Ltd., London, 164p.

Kao, Y.-hs. (ed.) 1962: Some Problems on Monsoons in East Asia. (in Chinese) Peking, 106p.

Kao, Y.-hs., Hsü, Sh.-y. 1962: Onset and retreat of monsoon and duration of the rainy season in East Asia. *In* Some Problems on Monsoons in East Asia (Kao, *ed.*). 78–87.

Kendrew, W. G. 1927: The Climates of the Continents. Oxford Univ. Press, Oxford, 400p.

Knoch, K., Schulze, A. 1954: Methoden der Klimaklassifikation. *Peterm. Mitt. Ergänzungsheft* (249) 1–78.

Kobayashi, T., Shikama, T. 1961: The climatic history of the Far East. *In* Descriptive Palaeoclimatology. (Nairn, A. E. M., ed.). New York, 380p., 292–306.

Köppen, W. 1936: Das geographische System der Klimate. *In* Handbuch der Klimatologie. Band I, Teil C, 1–44.

Kuo, Ch.-y., Yang, Ch.-sh. 1965: State and structure of the atmospheric circulation in midsummer in China. *Tiri-chi-kan* (9) 69–84.

Kurashima, A. 1959: Taiki kanryū to kisetsufū (General circulation and monsoon). *In* Kisetsufū (Monsoons) by Nemoto, Kurashima, Yoshino, and Numata. 201–283.

Kurashima, A. 1968: Studies on the winter and summer monsoons in East Asia based on synoptic and dynamic concept. *Geoph. Mag.* **34**(2) 145–235.

Lamb. H. H. 1961: Fundamentals of climate. *In* Descriptive Palaeoclimatology (Nairn, A. E. M., *ed.*). New York, 380p., 8–44.

Lang., R. 1915: Versuch einer exakten Klassifikation der Böden in klimatischer und geologischer Hinsicht. *Inter. Mitt. f. Bodenkunde* 1915 **5** 312–346.

Lang, R. 1920: Verwitterung und Bodenbildung: eine Einführung in die Bodenkunde. Stuttgart, 188S.

Lu, Ch.-y. *et. al.* 1965: Study on the wet and dry periods and regionalization of China according to aridity. *Acta Geogr. Sinica* **31** 15–24.

Maejima, I. 1967: Natural seasons and weather singularities in Japan. *Tokyo Metropolitan Univ., Geogr. Rep.* (2) 77–103.

Mallik, A. K. 1958: Is the incidence of droughts increasing in Bihar and eastern districts of Uttar Pradesh? *In* Proc. Symp. Met. Hydr. Aspects of Floods and Drought in India. India Met. Dept. 65–71.

Martonne, E. de 1926: Une nouvelle function climatologique: L'indice d'aridité. *La Met.* 449–458.

Meyer, A. 1926: Über einige Zusammenhänge zwischen Klima und Boden in Europa. *Chemie der Erde* (2) 209–347.

Murakami, T. 1959: The general circulation and water-vapour balance over the Far East during the rainy season. *Geoph. Mag.* **29** 131–171.

Naqvi, S. M. 1949: Coefficient of variability of monsoon rainfall in India and Pakistan. *Pakistan Geogr. Rev.* **4**(2) 7–17.

Nishiguchi, T. 1968: On the recent status of irrigation and rainfall distribution characteristics in the Philippines. *Southeast Asian Studies* **5**(4) 804–807.

Ogasawara, K. 1945: Nanpō Kikōron (Climatology of the South). Sansei-do Book Co., Tokyo, 396p.

Ogino, K. 1967: A climatological classification of Thailand with special reference to humidity. *Southeast Asian Studies* **5** (3) 500–530.

Passerat, C. 1906: Les pluies de mousson en Asie. *Ann. Geogr.* **15** 193–212.

Prescott, J. A. 1958: Climatic indices in relation to the water balance. *In* Climatology and microclimatology (UNESCO). 48–51.

Raman, P. K., Krishnan, A. 1960: Runs of dry and wet spells during southwest monsoon and onset of monsoon along the west coast of India. *Ind. J. Met. Geoph.* **11** 105–116.

Sadler, J. C. 1969: Average cloudiness in the tropics from satellite observations. East-West Center Press, Honolulu, 23p.+Figs.

Saito, N. 1966: A preliminary study of the summer monsoon of southern and eastern Asia. *J. Met. Soc. Japan* **44** 44–59.

Sawada, T. 1968: Large scale irrigation projects in Thailand. *Southeast Asian Studies* **5**(4) 786–792.

Schick, M. 1953: Die geographische Verbreitung des Monsuns. *Nova Acta Leopoldina N. F.* **16**(112) 127–257.

Schweinfurth, U., Flohn, H., Domrös, M. 1970: Studies in the Climatology of South Asia. Franz Steiner Verl., Wiesbaden, 16p.+15 Maps.

Shen, Ch.-ch. 1965: Three-dimensional structure of the prolonged dry period in the Kiang Hwai region in July, 1959. *Tiri-chi-kan* (9) 19–32.

Sion, J. 1928: Asie des moussons. Libr. Armand Colin, Paris, 272p.

Sircar, P. K. 1958: Variability of the annual rainfall of India. *Geographer* (Muslim Univ. Aligarb) **10** 1–9.

Staff Members, Acad. Sinica, Peking 1957, 1958a and b: On the general circulation over Eastern Asia (1), (2), (3). *Tellus* 9 432–446, 10 58–75, 10 299–312.

Stamp, L. D. 1962: Asia. A regional and economic geography. Dutton, New York, 730p.

Starr, V. P., Peixoto, J. P. 1958: On the global balance of water vapour and the hydrology of deserts. *Tellus* 10 189–194.

Starr, V. P., Peixoto, J. P. 1964: The hemispheric eddy flux of water vapour and its implications for the mechanics of the general circulation. *Arch. Met. Geoph. Biokl. (A)* 14(2) 111–130.

Starr, V. P., Peixoto, J. P., Crisi, A. R. 1965: Hemispheric water balance for the IGY. *Tellus* 17(4) 463–472.

Subrahmanyam, V. P. 1958: Droughts and aridity in India. A climatic study. *In* Proc. Symp. Met. and Hydr. Aspects of Floods and Droughts in India, India Met. Dept., 171–177.

Tada, F. 1938: Über die periodische Änderung der Regenmenge in Chosen seit dem Jahre 1776. Comptes Rendus du Congr. Intern. Géogr. Amsterdam 1938, 2 A-F 305–308.

Thompson, B. W. 1951: An essay on the general circulation of the atmosphere over Southeast Asia and the West Pacific. *Q. J. Roy. Met. Soc.* 77 569–597.

Thornthwaite, C. W. 1931: The climate of North America according to a new classification. *Geogr. Rev.* 21 633–655.

Thornthwaite, C. W. 1933: The climate of the earth. *Geogr. Rev.* 23 433–440.

Thornthwaite, C. W. 1948: An approach toward a rational classification of climate. *Geogr. Rev.* 38 55–94.

Trewartha, G. T. 1954: An Introduction to Climate. London, 402p.

Trewartha, G. T. 1962: The Earth's Problem Climates. Univ. Wisc. Press, Madison, 334p.

Tsuchiya, I. 1969: Selected bibliography on water balance of Monsoon Asia (1). *Climatological Notes, Hosei Univ.* (3) 1–28.

Wada, H. 1962: A study on the behaviours of the polar vortex and its application to long range weather forecasting. *Pap. Met. Geoph.* 31(2).

Wang, Z.-sh., Xu, X. 1964: An investigation on the water cycle in the basin of the upper-Yangtze river in summer. *Acta Met. Sinica* 34(3) 345–354.

Watts, I. E. M. 1955: Equatorial Weather. London, 224p.

Watts, I. E. M. 1969: Climates of China and Korea. *World Survey of Climatology* 8 1–117.

Woeikov, A. 1887: Die Klimate der Erde. Teil II. 422 S.

Yamamoto, T. 1952: Secular variation of summer rainfall in the Far East. *Kagaku* 22 96 (Transl. Directorate Sci. Inf. Serv. DRB Canada 1959 T66J).

Yamamoto, T. 1958: On the mechanism of the climatic changes in Japan. *Geoph. Mag.* 28 505–515.

Yao, Ch.-sh. 1958: The variability of precipitation in Eastern China. *Acta Met. Sinica* 29 225–238.

Yao, Ch.-sh. 1965: Climatic Statistics. Peking, 246p.

Yoshino, M. M. 1963: Rainfall, frontal zones and jet streams in early summer over East Asia. *Bonner Met. Abhandl.* (3) 1–127.

Yoshino, M. M. 1965a: Frontal zones and precipitation distribution in the rainy season over East Asia. *Geogr. Rev. Japan* 38(1) 14–28.

Yoshino, M. M. 1965b, 1966: Four stages of the rainy season in early summer over East Asia (Part I). *J. Met. Soc. Japan* 43 231–245, (Part II) 44 209–217.

PART II

WATER BALANCE AND ATMOSPHERIC CIRCULATION
OVER MONSOON ASIA

TRANSPORT AND SOURCE OF WATER VAPOR
IN THE NORTHERN HEMISPHERE AND MONSOON ASIA

Tadashi ASAKURA

Abstract: The precipitable water in Monsoon Asia reaches nearly 20%
of its total in the Northern Hemisphere, ranging from 19.1% in January to
28.2% in July. The mean value of the precipitable water over Monsoon
Asia is larger than that over the Northern Hemisphere during the summer
half of the year, especially in July; the former amounts to as much as
127% of the latter. Firstly, transport of water vapor was analyzed in the
Northern Hemisphere and Monsoon Asia. Water vapor transport in general originates from the subtropical anticyclone over the oceans and
grows larger along the outer boundary of the subtropical anticyclone.
Origins of water vapor transport were found around the Southwest
Pacific Ocean, east of the Philippines and the Atlantic Ocean in winter
and the Central Pacific Ocean, east of the Ryukyus, the Arabian Sea,
the Bay of Bengal and the Atlantic Ocean in summer. These areas are the
sources of water vapor which is supplied from the ocean surface. In
July, one sources of water vapor is located over China, suggesting the
possibility of evaporation from the land. Secondly, the northward transport of water vapor by meridional circulation and eddies was analyzed.
The southward transport of water vapor is found in the polar region only
in summer, suggesting evaporation from the open sea of the Arctic Ocean.

1. INTRODUCTION

Studies on the transport of water vapor in Asia have been made mainly for the summer
monsoon. Over India, Thompson (1951) pointed out that the Indian westerlies start west of
India and flow into South China, sometimes into southern Japan. Dao and Chen (1957)
found that the air mass of Indian westerlies was not of tropical but of subtropical origin.

Flohn (1957) suggested that in humid regions the evaporation over land is only slightly
smaller than oceanic evaporation and that the transport of water vapor from the continent
is large in East Asia. Murakami (1959) also discussed the transport of water vapor in the
Baiu season (the rainy season in early summer in Japan), 1957, and found many interesting
facts: For example, water vapor transport is to a large extent controlled by the prevailing
currents, one being controlled by the Indian westerlies centered at about the 700 mb level
and the other by the easterlies close to the ground surface. The former moisture flow plays
an important role in the early stage of the summer monsoon rain in East Asia and the latter
is significant in the last stage. Estimating the amount of evaporation from precipitation data
and the individual change of the mixing ratio integrated with respect to pressure, he found
that the evaporation is intense east of Taiwan.

Saito (1966) further analyzed the transport of water vapor in summer in 1964 and pointed
out that the wet tongue is found usually north of the SW or SE monsoon rather than in the

monsoon itself. The moist air on the continent is associated with the convergence between the southerly flow and the tropical continental air in China. Over Japan and the neighboring seas, the net inflow of vapor from the south is larger than the net outflow of vapor.

These studies used data over East Asia in June and July. In this paper, we used data over the Northern Hemisphere for one year (Oct. 1968 to Sept. 1969) and discussed the transport of water vapor and its source regions.

The data sources used in this study are the automatic computer-analyzed data based on weather messages in the Northern Hemisphere. The isobaric height, temperature, dew-point temperature and wind at 1,000, 850, 700, 500, 300, 200, and 100 mb levels at each grid point (grid size: 381 km) were available. The accuracy of the data in the low latitudes is poor, however, because of fewer weather messages.

2.   THE PRECIPITABLE WATER IN ASIA AND IN THE NORTHERN HEMISPHERE

The precipitable water in Asia, East Asia, and the Northern Hemisphere was calculated for five days in each month for one year. The regions of calculation are shown in Fig. 1 and the results are tabulated in Table I.

In the Northern Hemisphere, the amount of precipitable water is largest in summer, reaching to about 130% of the annual mean, while it is only 75% in February. The average of the precipitable water in Asia is 22.5% of the total for the Northern Hemisphere. The maximum concentration, 28.2% is found in July when the summer monsoon is at its peak and the minimum, 19.1–19.5% occurs in November through January when the winter monsoon prevails. In East Asia, the average of precipitable water is 10.7% of the total for the Northern Hemisphere. The maximum concentration, 12.5% occurs also in July and the minimum, 8.6%, in November.

Table I.   Precipitable water in the Northern Hemisphere, Asia and East Asia. The values in Asia and East Asia are presented by percentages to total in the Northern Hemisphere.

Date	Northern Hemisphere (g/kg) · mb · 381×381 km²	Asia, %	East Asia, %
Oct. 14–18	15,295.5	20.4	10.3
Nov. 11–15	12,912.7	19.5	8.6
Dec. 16–20	11,411.6	19.4	9.0
Jan. 19–23	11,071.0	19.1	10.1
Feb. 16–20	10,784.9	21.0	10.3
March 23–27	11,780.4	24.1	11.9
Apr. 21–25	13,979.9	22.6	10.9
May 26–30	15,613.3	24.8	11.6
July 7–11	18,738.1	28.2	12.5
Sept. 1–5	19,091.4	25.4	12.1
Sept. 29–Oct. 3	17,005.7	22.0	10.5

However, from these percentages, one cannot say that Asia is relatively wet. To compare the degree of wetness in different regions, the space means of the precipitable water were calculated as shown in Fig. 2, on the basis of Table I. Throughout the year, East Asia is the wettest, especially in June and July. For example, the mean precipitable water in each zone in July is 19.4 mb g/kg (East Asia), 14.3 mb g/kg (Asia) and 11.3 mb g/kg (N.H.). The former is 1.71 times the latter. It is drier in Asia than in the Northern Hemisphere during the winter because of the predominance of cool and dry weather in Siberia, while in summer, the

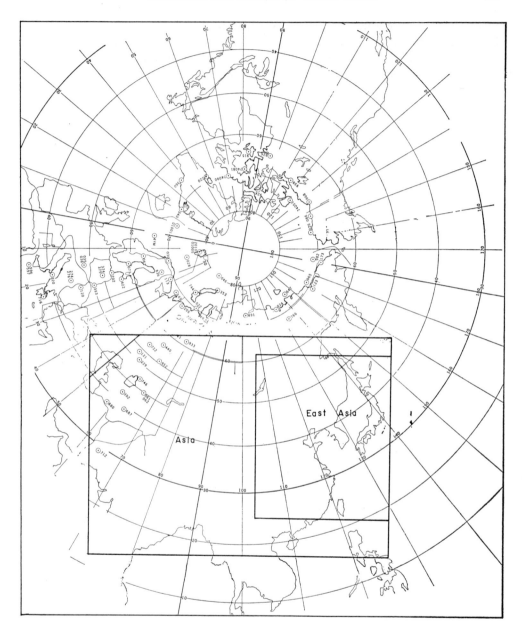

Fig. 1. The regions of the Northern Hemisphere, Asia and East Asia used in the present calculation.

amount of precipitable water in Asia is larger than in the Northern Hemisphere.

In summary, it may be concluded that Monsoon Asia in summer is characterized by very wet weather which is 1.7 times wetter than the mean value for the Northern Hemisphere. On the other hand, Monsoon Asia in winter is drier than the Northern Hemisphere.

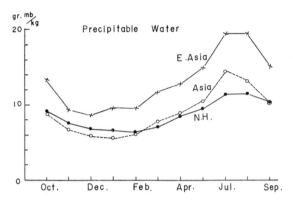

Fig. 2. Five day mean precipitable water in the Northern Hemisphere, Asia and East Asia for each month in the year (Oct. 1968 to Sept. 1969).

3. INFLOW AND OUTFLOW OF WATER VAPOR OVER ASIA AND EAST ASIA

The inflow and outflow of water vapor over Asia and East Asia was evaluated for eleven months. The region of Asia and East Asia is defined as shown in Fig. 1. Over Asia, the length has 16 grids (381×16 km) and the width 23 grids (381×23 km). Over East Asia, the length has 12 grids (381×12 km) and the width 10 grids (381×10 km).

The inflow and outflow of water vapor was integrated with respect to pressure between 1,000 and 500 mb. That is, the flow of water vapor passing through a pressure wall of 1,000–500 mb was evaluated. Results are tabulated in Tables II and III for Asia and East Asia, respectively.

The first column in the tables shows the period of calculation, the second the convergence of water vapor east-westward (that is, the outflow along the east length minus the inflow along the west length), the third the convergence of water vapor north-southward (that is,

Table II. Convergence of water vapor in East Asia (g/kg · m/sec · mb · 381 km) (1968–1969).

Date	East-West	North-South	Total
Oct. 14–18	582.0	4,630.0	5,212.0
Nov. 11–15	1,621.2	2,373.0	3,994.2
Dec. 12–16	1,875.6	567.0	2,442.6
Jan. 19–24	3,297.6	−2,283.0	1,014.6
Feb. 16–21	2,720.4	−2,862.0	−141.6
March 23–27	2,822.4	−2,084.0	738.4
Apr. 21–25	1,078.8	−1,472.0	−393.2
May 26–30	1,311.6	−3,165.0	−1,853.4
July 7–11	3,652.8	−3,655.0	−2.2
Sept. 1–5	−3,517.2	3,254.0	−263.2
Sept. 29–Oct. 3	−2,828.4	−572.0	−3,400.4

the outflow along the north width minus the inflow along the south width), the fourth the sum of the two convergences.

As shown in Table II, the flow of water vapor diverges from October to January. The divergence of water vapor in October is largest and is about twice that in January. The divergence of water vapor brings dry weather to Asia, the driest being in January. As the tendency of water vapor is generally negligible, the divergence of water vapor is the result of a deficit in precipitation more than evaporation. The continuous deficit of precipitation ends in January, resulting in the driest month.

Table III. Convergence of water vapor in Monsoon Asia (g/kg · m/sec · mb · 381km) (1968–1969).

Date	East-West	North-South	Total
Oct. 14–18	3,404.8	7,341.6	10,746.4
Nov. 11–15	2,374.4	5,881.1	8,255.5
Dec. 12–16	3,273.6	5,529.2	8,802.8
Jan. 19–24	4,113.6	105.8	4,219.4
Feb. 16–21	1,476.8	−1,596.2	−119.4
March 23–27	1,470.4	−2,249.4	−779.0
Apr. 21–25	−1,611.2	365.7	−1,245.5
May 26–30	1,510.4	−6,975.9	−5,465.5
July 7–11	−5,416.0	−4,618.4	−10,034.4
Sept. 1–5	−14,123.2	7,725.6	−6,597.6
Sept. 29–Oct. 3	820.8	4,117.0	4,937.8

In February and March, the water vapor diverges east-westwards while it converges north-southwards; in total, it converges. During the summer, the convergence of water vapor increases with the season and attains its maximum value in July, when the summer monsoon is at its peak. The convergence of water vapor over Asia starts in February and ends in September. The value of convergence in February is only about 1 % of that July. A sizable increment in convergence takes place in May and July and the convergence in the latter month is about twice that in the former.

In East Asia, a divergence of water vapor is also observed from October to January as shown in Table III. The maximum divergence occurs in October; after that its value decreases, and finally in February the water vapor converges. From February to September, the water vapor converges, except in March. From March 23–27, 1969, there occurred a revival of the cold air outbreak over East Asia and therefore the water vapor diverged there as in winter. The water vapor converged intensely at the end of May, when the summer monsoon began and converged weakly from July 7–11, when the summer monsoon was coming to an end. Another large convergence of water vapor was found from September 29 to October 3, during which time a typhoon was in the area.

4. THE MERIDIONAL TRANSPORT OF WATER VAPOR

The meridional transport of water vapor is calculated from wind observations. The transport of water vapor over the region to the south of 20° latitude has not been examined in detail because of the scarcity of data. The calculations are for five days monthly from October 1968 to September 1969. The annual mean is the average of the period described above.

The meridional transport of water vapor is carried out by meridional circulation and eddies. In this calculation, the order of transport of water vapor by meridional circulation is larger than that by eddies.

The annual mean of the meridional transport of water vapor is directed northward over the larger part of the area north of 30°N and southward over the low latitudes as shown in Fig. 3. The center of the northward transport appears at the 850 mb level around 35–40°N, and the center of the southward transport is above the 700 mb level.

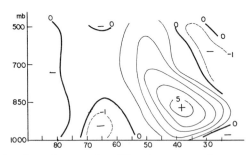

Fig. 3.   Annual mean of meridional transport of water vapor.

As mentioned earlier, the meridional transport of the water vapor consists of the meridional circulation and eddies. The transport of water vapor by meridional circulation shows a three cell pattern as indicated in Fig. 4a: the southward transport is in the high and low latitudes and the northward transport in the middle latitudes. The center of the northward transport is located at the 850 mb level at 35°N. The centers of the southward transport are at the 1,000 mb level at 65°N and at 25°N or so. The former seems to correspond to the polar frontal zone and the latter to the subtropical anticyclone.

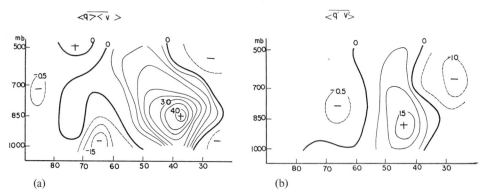

(a)   (b)

Fig. 4.   Annual mean of meridional transport of water vapor by meridional circulation (a) and by eddies (b).

The meridional transport of water vapor by eddies shows a similar pattern to that of the meridional circulation as shown in Fig. 4b. Some differences, however, exist: The center of northward transport of water vapor by eddies is 5° latitude further north than the center of meridional circulation and the center of southward transport by eddies is at the 700 mb level at 25°N. As for the amount of transport, the southward eddy trnasport in the low latitudes is stronger than the meridional transport, but the northward eddy transport in the middle latitudes is weaker.

These features change with the seasons, however, as shown in Fig. 5. In December, transport by meridional circulation consists of four cells: northward transport in the polar region, southward transport in the high latitudes, northward transport in the middle latitudes and southward transport in the low latitudes. In July, transport by meridional circulation consists of three cells which have nearly the same mode of transport as that of the annual mean. In the polar region, northward transport in December changes to southward transport in July, while in the high latitudes, southward transport in December changes to northward transport in July. Southward transport in the low latitudes is stronger in July than in December.

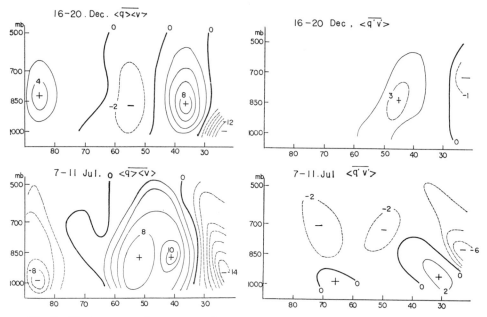

Fig. 5. The transport of water vapor by meridional circulation and eddies for December 16–20, 1968 and for July 7–11, 1969.

The meridional transport by eddies is relatively less than the transport by meridional circulation. The transport by eddies is directed northward over the middle and high latitudes and southward over the low latitudes in December. In July, on the other hand, transport by eddies is directed southward over the region south of 30°. The maximum northward transport is at the 850 mb level. The maximum southward transport by meridional circulation is located at the 1,000 mb level both in December and July.

The total meridional transport of water vapor integrated between 1,000 and 500 mb for the annual mean in January, April, July, and October is shown in Fig. 6. In this figure, the full line shows transport by meridional circulation, the dotted line transport by eddies. For the annual mean, the northward transport by meridional circulation is shown, although a small value is found for southward transport around 60°N as well as southward transport north of 75°N. The transport by eddies in the middle and low latitudes is strong and in the former is directed northward while in the latter southward. It is very weak at high latitudes however. In total, the water vapor is transported polewards in the middle and high latitudes and equatorwards in the low latitudes. Accordingly, the source of the water vapor is to be found at 25°N.

In January, the signs of transport by the meridional circulation cell and eddies are reversed in the middle and high latitudes. Therefore, the total is very small except in the middle latitudes. The eddy transport is very active in the middle latitudes and is much larger than transport by the meridional circulation cell. The eddy transport is less active in April than in January, while transport by the meridional circulation cell is more active. The meridional circulation shows a pronounced four-cell structure; that is, the transport is directed northward in the polar region, southward in the high latitudes, northward in the middle latitudes and southward in the low latitudes. The water vapor is transported southward by the eddies over the region south of 40°N and northward over the region between

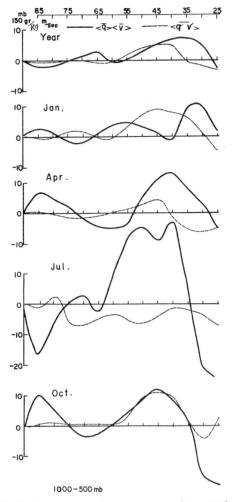

Fig. 6. Transport of water vapor integrated with respect to pressure between 1,000–500 mb for January, April, July, and October of one year. Full line: meridional circulation; dotted line: eddies.

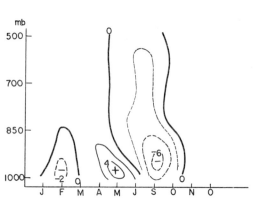

Fig. 7. Water vapor transport at 75°N lat. for each month (Oct. 1968 to Sept. 1969).

40°N and 60°N. In total, the source of the water vapor is to be found around 35°N and the sink around 55°N.

In July, the eddy transport is rather weak in comparison with the other months and the cell transport is very active, forming a three-cell structure. The northward transport of water vapor is found over the region between 35–75°N and the southward transport over the region south of 35°N. The source of water vapor is, therefore, to be found around 35°N. Another source of water vapor is found around the polar region.

In October, the eddy transport is active only in the middle latitudes and very weak in other latitudes. Transport by the meridional circulation cell is nearly the same as in April.

It is interesting to note that the southward transport of water vapor takes place in the polar region in July. As shown in Fig. 7, the transport of water vapor crossing the 75°N circle shows a seasonal change. A southward flow of water vapor from the Arctic Ocean is observed during the summer from May to September. During the winter, however, a northward inflow of water vapor to the Arctic Ocean is generally observed, except in February. These facts suggest that evaporation may take place from the open sea of the Arctic Ocean.

5.   THE SOURCES AND SINKS OF WATER VAPOR IN THE NORTHERN HEMISPHERE

The total transport of water vapor between 1,000 and 500 mb is calculated by using wind and specific humidity data at 1,000, 850, 700, and 500 mb levels. The source of water vapor at each grid point is estimated by the dynamical method. As stated in section 3, the grid size is 381 km and the data are for five days in each month from October 1968 to September 1969.

The equation for the specific humidity per unit volume may be written

$$\frac{\partial q}{\partial t} + \nabla(q\mathbf{V}) + \frac{\partial}{\partial p}(q\omega) = C + E, \tag{1}$$

where q is the specific humidity, $\mathbf{V}$ the horizontal velocity, ω the vertical p-velocity, C the rate of condensation, and E the rate of evaporation.

Integrating Eq. (1) with respect to pressure between 1,000 and 500 mb, we get

$$\int_{500}^{1000} \frac{\partial q}{\partial t}dp + \int_{500}^{1000} \nabla(q\mathbf{V})dp + (q\omega)_{1000} - (q\omega)_{500} = \int_{500}^{1000}(C+E)dp. \tag{2}$$

In Eq. (2), the third term is zero as $\omega = 0$ at 1,000 mb and the fourth term is nearly zero as q at 500 mb is very small; therefore the ω term in Eq. (2) can safely be neglected in this calculation. Equation (2) is then simplified as follows:

$$\int_{500}^{1000} \frac{\partial q}{\partial t}dp + \nabla \int_{500}^{1000} q\mathbf{V}dp = \int_{500}^{1000}(C+E)dp. \tag{3}$$

In this study, the left hand side of Eq. (3) is applied to a daily data and the time difference is 24 hours. Results of Eq. (3) at each grid point are averaged for five days in each month. A positive value means that the amount of evaporation is larger than the amount of condensation, a negative value, smaller. In other words, the former corresponds to the source, the latter to the sink of water vapor in the atmosphere.

The results, shown in Figs. 8a–j, are summarized as follows. The arrow in the figures indicates the direction of water vapor transport and the length is proportional to the amount of water vapor transport.

a. *January 19–24, 1969* (Fig. 8a)

The strong flow of water vapor is over the Pacific Ocean and the Atlantic Ocean and is not found over the continents. Over the Pacific Ocean, the second flow is along the polar front running from South China to the Aleutian Islands, and the third near the west coast of North America. A zone with wet flows runs over the Atlantic Ocean to the west coast of Europe.

The source regions of the water vapor are located over the Pacific Ocean with a center near the Central Pacific Ocean, and over the Atlantic Ocean with a center near the east coast of North America. The sinks of the water vapor appear over the Aleutian Islands, south of Iceland, Central China and the east coast of Japan. The former two correspond to precipitation from the Aleutian low and the Iceland low respectively and the latter two to precipitation by the cyclones on the polar front.

In Monsoon Asia, the main sources of water vapor are seen over the Pacific Ocean, the East China Sea, the Japan Sea and South China. On the other hand, the sinks of water vapor are found over the cyclone family along the polar front originating from the Aleutian low.

b. *February 16–21, 1969* (Fig. 8b)

There are two strong flows of water vapor, the one easterly and the other westerly. The former runs from the Mariana Islands to Vietnam passing over the Philippine Islands. The latter runs south of the Tibetan Plateau, north of Vietnam, the Ryukyus, the southern coast of Japan and the west coast of North America. There is no strong wet flow over the Atlantic Ocean during this period, except the narrow zone reaching to the west coast of the Iberian Peninsula.

The sources of water vapor are mainly over the Pacific Ocean and the Atlantic Ocean. Strictly speaking, the source regions are found west of California, the Central Pacific, east of the Philippines and Yunnan. The first two convergence flows of water vapor form a wet flow which reaches south of Alaska. The latter two convergence flows of water vapor join over South China, Taiwan, and the south coast of Japan just along the polar front, which remained stationary over the analyzed period. Over the Atlantic Ocean, there is a large divergence area of water vapor, with its center south of the Azores.

The sinks of water vapor are located over its convergence zone. Their centers are located over South China, Taiwan, and south of Japan and on the southern coast of Alaska. These coincide with the position of the polar front and with the paths of the Aleutian low along the arctic front. Another sink of water vapor is located over Burma and East Pakistan.

An elongated wide zone of the sink is located from the west coast of the Iberian Peninsula to the Persian Gulf, passing through the Mediterranean Sea and the Black Sea. This sink zone is nearly along the polar frontal zone in the Mediterranean Sea, where the precipitation is greater than the evaporation.

c. *March 23–27, 1969* (Fig. 8c)

Strong flows of water vapor are found over the Pacific Ocean and Southeast Asia but not over the continents. Over the Pacific Ocean, there are three systems of wet flow. One is the wet flow of easterlies and westerlies around the Philippines, the second flow is around the Hawaiian Islands and the third is over the Ocean west of California. Over South Asia, though this is not shown here, the wet flow starts from the Bay of Bengal and invades India.

Another wet flow starts from the tropical Pacific Ocean and flows into Vietnam. Over the Atlantic Ocean, the wet flow is not as strong as over the Pacific Ocean.

The sources of water vapor are located north of the Hawaiian Islands, east of Japan and the Mariana Islands and over the South China Sea. Over Asia, the sources and sinks of water vapor are alternately distributed and have large values. Over the Atlantic Ocean, the source of water vapor covers a wide area. The sources over the continents are very weak compared to those over the Oceans.

The sinks of water vapor are located over Northwest India, Northeast India and Burma, Yunnan, Taiwan and its northeastern sea, the broad area from the Central Pacific to the Aleutian Islands and off the west coast of North America. Other sinks are located over eastern North America and the southern part of the Atlantic Ocean. A large convergence region of water vapor is located over Italy and Turkey.

d.   *April 20–25, 1969* (Fig. 8d)

Strong wet air flows from the Arabian Peninsula via the Arabian Sea and India to northern Burma. A part of its wet flow joins over South China with the southwest wet flow from the southwest Pacific. This wet, rather strong flow extends over East China, the Yellow Sea and the Japan Sea. Another wet flow is found over the Central Pacific, heading east and gradually curving northward. Over the Atlantic Ocean, the strong wet air flows over the Azores Islands and invades western Europe.

The sources of the water vapor are located over the Pacific Ocean, the Arabian Peninsula and the Mediterranean Sea. Over the Pacific Ocean, the source has many centers. Zones of intense evaporation from the sea surface are located northeast of the Hawaiian Islands, near Marcus Island, in the East China Sea and in the South China Sea. The sinks of water vapor are found over Unnan, Central China, and the Japan Sea, and these regions belong to the polar frontal zone. There is another sink of water vapor over India and Burma. Over Europe, there is a zonal band of sink which extends from the eastern Atlantic Ocean to the Black Sea. The source of water vapor is located over the Mediterranean Sea though it is the sink in winter. Over North America, the convergence of water vapor is located over the Gulf of Alaska and the Great Lakes.

e.   *May 25–30, 1969* (Fig. 8e)

Over Monsoon Asia, there is a rather wet air flow compared with the flow over the other regions. In particular, a strong wet stream flows over the Arabian Sea, the Bay of Bengal, Vietnam, the South China Sea, the Ryukyus and Japan under the influence of the summer monsoon. This wet flow extends further east, reaching North America. It is interesting to note that an outflow of water vapor from the Arctic Ocean is apparent.

The source areas of water vapor over the Pacific Ocean are smaller in May than in April. However, the sources of water vapor are located over the continent, such as China, Iran, Saudi Arabia and eastern North America. This fact suggests the importance of evaporation from the ground surface as well as the sea surface in May.

The sinks of water vapor are located along the Baiu front near Taiwan and off the south coast of Japan. Another sink is over Burma and Pakistan and may correspond to the summer monsoon rain. Over Europe, the sink of water vapor is found over the Mediterranean Sea, with its center near Italy and its source over the wide region of Turkey and the Ukraine. Over North America, sinks and sources appear alternately within narrow areas.

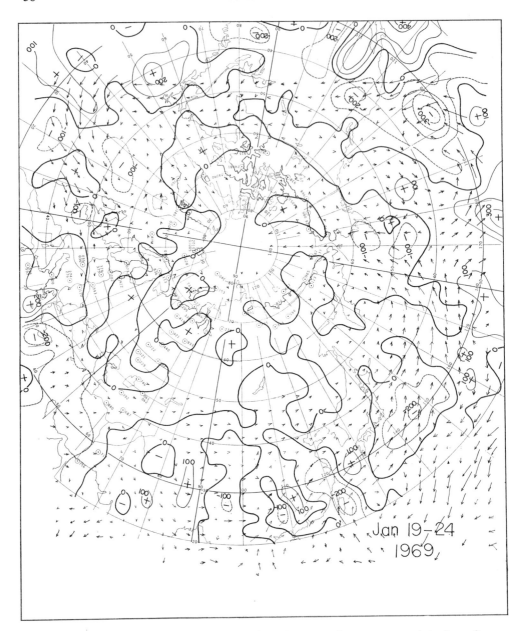

Fig. 8a. Distribution of 5-day mean of sinks and sources of water vapor and the flow of water vapor integrated between the 1,000 mb and 500 mb levels over the Northern Hemisphere for the period of January 19–24, 1969 (unit: $10^{-4} sec^{-1} g kg^{-1}$ mb). A plus sign means a source and a negative sign a sink of water vapor.

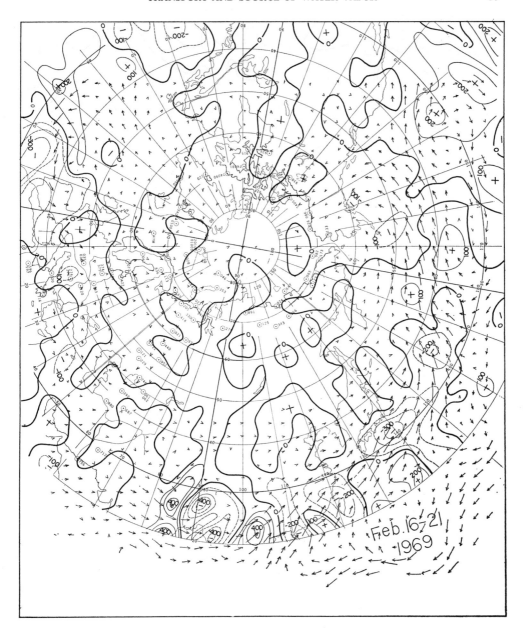

Fig. 8b. Same as Fig. 8a but for February 16–21, 1969.

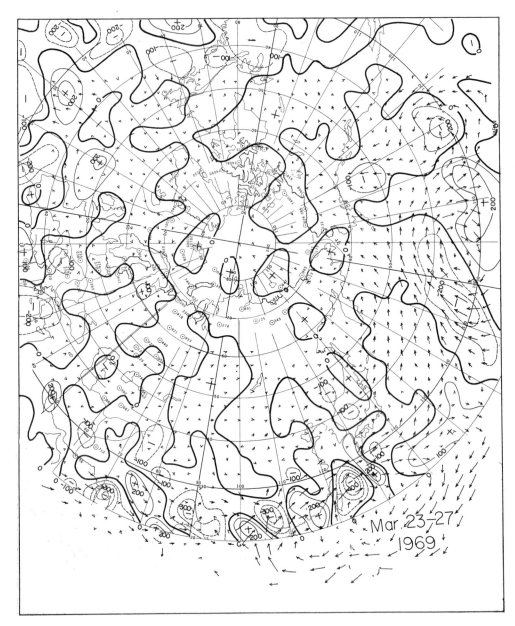

Fig. 8c. Same as Fig. 8a but for March 23–27, 1969.

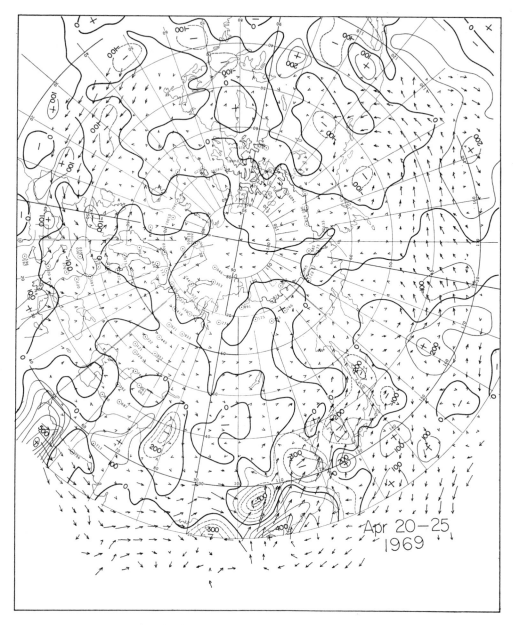

Fig. 8d. Same as Fig. 8a but for April 20–25, 1969.

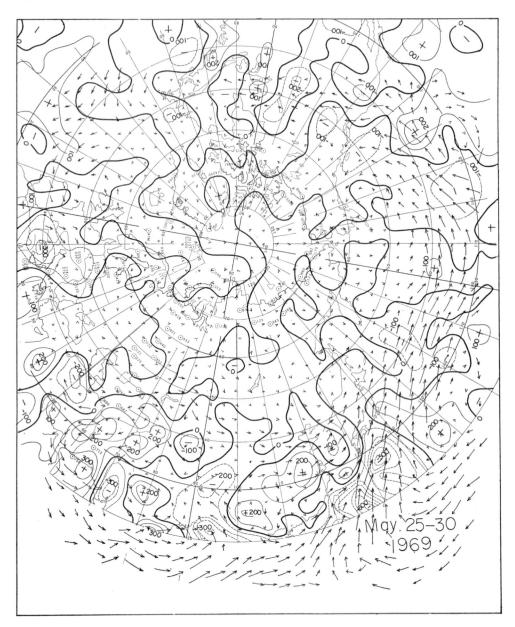

Fig. 8e.   Same as Fig. 8a but for May 25–30, 1969.

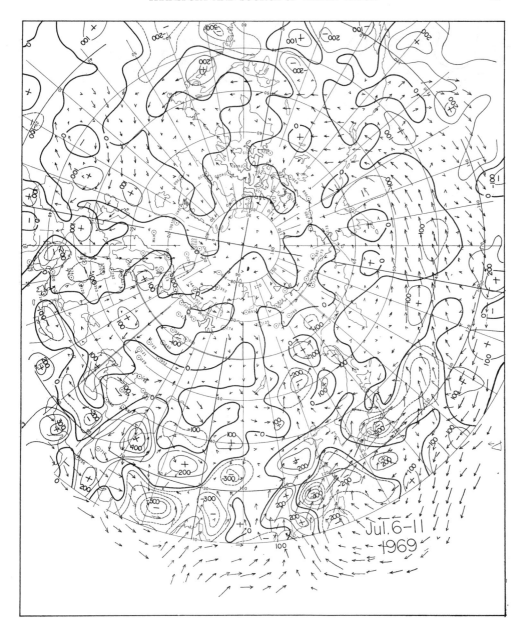

Fig. 8f. Same as Fig. 8a but for July 7–11, 1969.

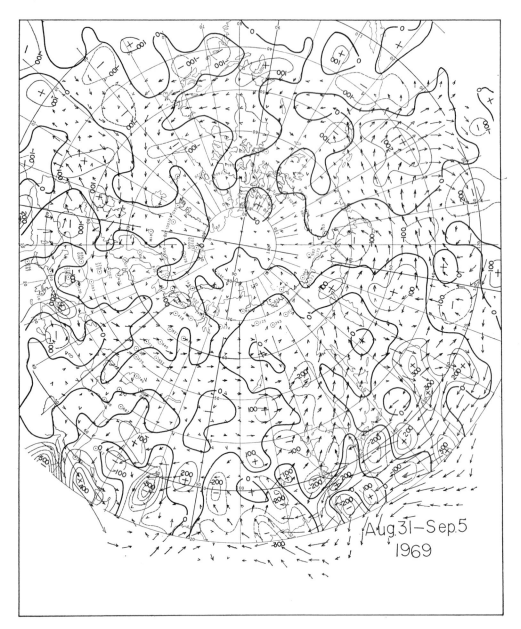

Fig. 8g.   Same as Fig. 8a but for August 31–September 5, 1969.

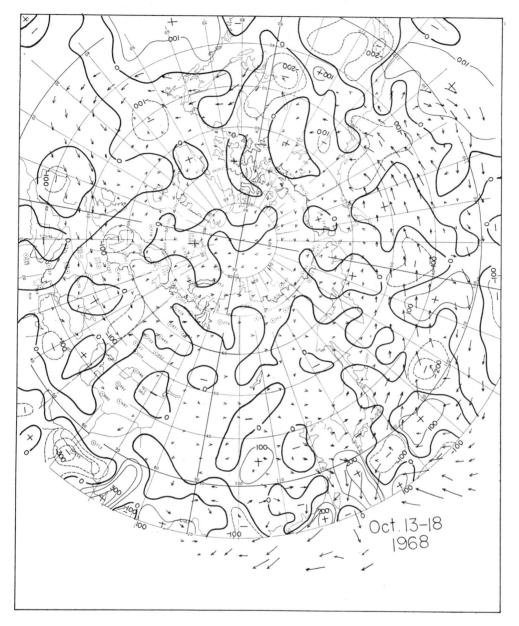

Fig. 8h. Same as Fig. 8a but for October 13–18, 1968.

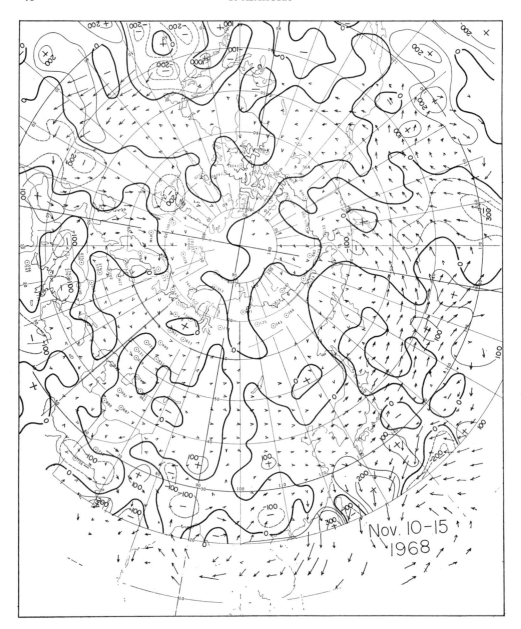

Fig. 8i. Same as Fig. 8a but for November 10–15, 1968.

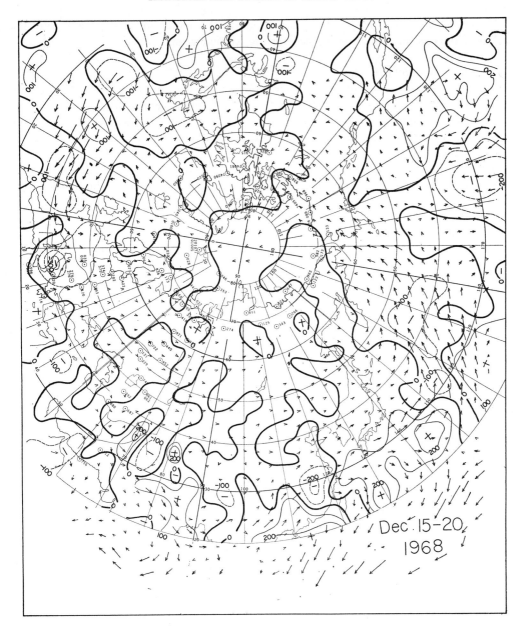

Fig. 8j. Same as Fig. 8a but for December 15–20, 1968.

As mentioned earlier, the peculiar phenomena for this month are the appearance of the sink over the continent and the source over the Arctic Ocean.

f. *July 7–11, 1969* (Fig. 8f)

This analyzed the period of the summer monsoon in Asia. As seen in Fig. 8f, the very strong, wet air flows stream with the easterlies and westerlies. Over South Asia, the strongest wet air flow is located over northern India, coming from the Arabian Sea. Another wet flow coming from the Bay of Bengal, invades Burma as the SW monsoon and part of it reaches Indochina, joining the wet southeasterlies over Vietnam.

Over the Pacific Ocean, a strong flow of wet air extends over the low latitudes: the northeasterlies over the Northwest Pacific, which turn to southeasterlies over the South China Sea. This flow occurs at the south and west margins of the subtropic anticyclones. North of the subtropic anticyclone there flows a strong southwesterly of very wet air with its center on the southern coast of Japan. This southwesterly originates from the South China Sea where the westerlies from the Bay of Bengal converge with the easterlies from the Pacific Ocean mentioned above.

Over the Atlantic Ocean, there is a strong flow of wet air which invades Europe. Over and around the Arctic Ocean, there is an inflow of water vapor near Scandinavia and an outflow near Siberia and Northern Canada. These two areas are the sources of water vapor.

The largest sources of water vapor are located over the Pacific Ocean east of about 160°E, the eastern part of China, the Takla Makan desert north of the Tibetan Plateau and Afghanistan. Evaporation from the latter two areas is relatively greater than from the Pacific Ocean.

Other sources are found over the Caspian Sea, Ukraine, and the eastern Atlantic Ocean. Special attention should be paid to the fact that the divergence of water vapor over the continent is larger than over the ocean.

The sinks of water vapor are found mainly over monsoon rain areas such as India, Burma, South China, the East China Sea and Japan. Another sink region appears south of the Aleutian Islands, over Canada and the west coast of the Atlantic Ocean.

g. *August 31 to Septembers 5, 1969* (Fig. 8g)

During this period, the tropical depression moves west across the Philippines and a strong wet air stream flows over the Pacific Ocean south of Japan and towards inland China. Part of it curves northeast and invades the Japan Sea. Another strong flow is located over the Pacific Ocean west of North America and over India from the Arabian Sea to the Bay of Bengal. Over the Atlantic Ocean, the wet air flows eastward towards the west of England, with its center near northern England.

The sources of the water vapor are found over the Pacific Ocean east of Japan and west of North America. The former is relatively strong and supplies water vapor to the flow around the tropical depression, while the latter supplies water vapor to the northeasterly flows. The sources of the water vapor near India are located over the Arabian Sea and the Indus in west Pakistan. The Black Sea and the Caspian Sea are also sources of water vapor. Over the Atlantic Ocean west of Spain there is a source which supplies water vapor to the inflows toward Europe. Over eastern North America, there is a source of water vapor with its center over the Great Lakes.

The sinks of water vapor in East Asia are found near the tropical depression east of the Philippines, China and Siberia. The latter corresponds to the frontal precipitation zone.

Over South Asia, two large sinks are found, one over South China and Burma, and the other over northern India. These regions belong to the summer monsoon zones.

Other sources over the Pacific Ocean are located over the Aleutian Islands, the Midway Islands and the Hawaiian Islands. Over the Atlantic Ocean east of North America and western Europe, additional sinks of water vapor are located, but these are not so extensive.

h.   *October 13–18, 1968* (Fig. 8h)

A very strong wet air flow is found over the Pacific Ocean east of the Philippines; part of its flow curves westward and eventually invades Vietnam. Then, in contrast to summer, wet easterlies flow over Indochina. In addition, wet easterlies flow over India from the Bay of Bengal.

Over the Pacific Ocean, wet westerlies flows from the east coast of Japan to the west coast of North America; their centers are located northeast of the Ogasawara Islands and the Hawaiian Islands. Over the Atlantic Ocean, wet westerlies flow from the Great Lakes of North America to Europe, crossing the Atlantic Ocean.

The sources of water vapor are mainly located over the Pacific Ocean: the East China Sea, around the Ogasawara Islands and around the Hawaiian Islands, and over the Atlantic Ocean east of North America. On the whole, the sources on the continents are weaker than in summer, except for India.

The sinks of water vapor over the Pacific Ocean are located east of Japan to the Aleutian Islands and south of the Gulf of Alaska. The former corresponds to the polar front. The sink region over India in the monsoon season retreats and is replaced by a source region, and a new sink region of water vapor appears over the Persian Gulf.

Over the Atlantic Ocean, a sink is found south of Greenland. This sink zone of water vapor continues from the United States to Europe, passing Iceland and Scandinavia.

i.   *November 11–15, 1968* (Fig. 8i)

The very wet air flows over South Asia and the Pacific Ocean are very similar to the conditions in October described in the previous section. But some differences are found north of the Hawaiian Islands where a strong wet southerly flow prevails.

Over South Asia, it is interesting to note that a strong wet air flows only over the Bay of Bengal towards the northwest while there is a very weak flow over India. Over the Atlantic Ocean, strong wet air flows west and curves northward off the west coast of Europe.

Sources of water vapor are found over the greater area of the Pacific Ocean; the centers are located over the South China Sea, the Ryukyus, the Central Pacific and the eastern Pacific west of California. A weak source is found over China and the Japan Sea, as well as over the Atlantic and the western Mediterranean Sea.

A strong sink of water vapor is located south of Alaska and a weak sink east of Japan. Another sink is found over Yunnan, Burma and the Persian Gulf. The latter sink region continues over the Caspian Sea, Turkey, and Italy. Over the Atlantic Ocean, two strong sinks are found off the east coast of North America and off the west coast of Europe.

j.   *December 15–20, 1968* (Fig. 8j)

Strong flows of wet air occur only over the oceans. Over the Pacific Ocean, a very wet northeasterly air flows over and around the South China Sea and the Philippines. The wet southwesterlies flow over the Northwest Pacific and the southerlies over the Northeast

Pacific north of the Hawaiian Islands. Wet air also flows over the Bay of Bengal and the Atlantic Ocean.

Large sources of water vapor are located over the ocean. In Asia, the sources are found over the South China Sea, the Ryukyus and the Ogasawara Islands. Other sources over the Pacific Ocean are found over the Central Pacific Ocean and off the west coast of North America. Over the Atlantic Ocean, there is another source of water vapor.

Large sinks of water vapor are located off the east coast of Japan and over the Aleutian Islands. These sinks correspond to precipitation from the polar front and the Aleutian low. Along the Yangtze, there also exists a sink of water vapor. Over India, no sink of water vapor is found except in summer but the source is located in West Pakistan.

A large sink region is located over the Mediterranean Sea with its center over Spain and Italy. This seems to correspond to the Mediterranean front. Another large sink is located over the east coast of North America.

6. CONCLUSION

This study dealt with the transport of water vapor, and the sources and sinks of water vapor in the Northern Hemisphere. The results are summarized as follows:

i.   In the Northern Hemisphere, the precipitable water is greatest in summer, amounting to about 130% of the annual mean, while it is only 75% of the annual mean in February.

ii.   East Asia is the wettest region throughout the year, especially in June and July.

iii.   The inflow and outflow of water vapor in Asia and East Asia show a pronounced seasonal change. During the summer, the convergence of water vapor increases with the passing of the season and attains a maximum value in July.

iv.   The meridional transport of water vapor is carried out by meridional circulation and eddies. The former shows a three-cell and sometimes a four-cell structure. The center of the northward transport is at the 850 mb level at 35°N. The centers of the southward transport are found at the 1,000 mb level at 65°N and at approximately 25°N. The southward eddy transport at low latitudes is stronger than the meridional transport, but the northward eddy transport at middle latitudes is weaker than the meridional transport.

v.   A southward flow of water vapor from the Arctic Ocean is observed during the summer from May to September. This suggests the possibility of evaporation from the open sea of the Arctic Ocean.

vi.   The sources and sinks of water vapor were studied from aerological data and the continuity equation of water vapor. In general, the sources of water vapor are found over the oceans where the centers of subtropical anticyclones form. There are also sources of water vapor over the continents during the summer, suggesting evaporation from the land surface.

ACKNOWLEDGEMENTS

The author wishes to thank Prof. M. M. Yoshino, Dr. H. Wada, and the other members of the research group on the water balance of Monsoon Asia for their valuable discussions and suggestions. The author is also indebted to Mr. M. Matsushita and Miss R. Shitihyo for their help in preparing this manuscript.

References

Benton, G. S., Estoque, M. A. 1954: Water vapour transfer over the North American Continent. *J. Met.* **11** 462–477.

Dao, Shin-Yen, Chen, Lung-Shun 1957: The structure of general circulation of Asia in summer. *75th Anniversary Volume of J. Met. Soc. Japan* 215–229.

Flohn, H., Oeckel, H. 1957: Water vapour flux during summer rains in Japan and Korea. *Geoph. Mag.* **27** 527–532.

Flohn, H. 1957: Large scale aspects of the "Summer-Monsoon" in South and East Asia. *75th Anniversary Volume of J. Met. Soc. Japan* 180–186.

Kurihara, Y. 1959: Consideration on the heat and water vapor budgets over U.S.A. in January 1957. *J. Met. Soc. Japan*, 258–273.

Murakami, T. 1959: The general circulation and water-vapor balance over the Far East during the rainy season. *Geoph. Mag.* **29** 131–171.

Saito, N. 1966: A preliminary study of the summer monsoon of southern and eastern Asia. *J. Met. Soc. Japan* **44** 44–59.

Samuel, B. S. 1939: Computation of depth of precipitable water in a column of air. *Mon. Wea. Rev.* **67** 100–103.

Santon, E. T. 1968: World distribution of mean monthly and annual precipitable water. *Mon. Wea. Rev.* **96** 785–797.

SEASONAL VARIATION IN WATER VAPOR BALANCE
OVER THE NORTHERN HEMISPHERE AND ASIA

Isao Kubota

Abstract: The seasonal variation in the horizontal divergence of the vertically integrated water vapor flux over the Northern Hemisphere and Asia is discussed here. The hydrological method was applied to compute the amount of divergence of the vertically integrated water vapor flux, substituting the amount of evaporation estimated by Budyko (1963), the precipitation and the time change of precipitable water estimated by the author into the water vapor balance equation. It was confirmed that the possible range of variation over the Hemisphere of the time change term of precipitable water throughout the year is too small to affect the seasonal variation of the water vapor divergence term. The hemispheric distributions of water vapor divergence for every month show that in the winter monsoon season, the source region appears most systematically and intensively over the subtropical western Pacific and the sink region over Malaysia; in the summer monsoon season, the source region is found over the Red Sea, the Caspian Sea, the Persian Gulf, the Arabian Sea, and the sink region on the continental coast over Southeast Asia. Asia. With respect to zonal averages, it was shown that the water vapor is transferred northward at all latitudes in the winter half-year. An especially strong tendency is found near 50°N and 10°N. In contrast, the water vapor is transferred over the tropics and northward over the higher latitudes in summer.

1. INTRODUCTION

There are two methods for evaluating the water vapor balance of the atmosphere, the aerological and the hydrological method. The water vapor balance equation of the atmosphere consists of the following four terms:

(a)   local time change of precipitable water;
(b)   horizontal divergence of vertically integrated water vapor flux;
(c)   evaporation from the surface; and
(d)   precipitation.

In the aerological method, (b) is computed from aerological data, and (c) and (d) are estimated. The term (a) is negligible over a relatively long period. In the hydrological method, (c) and (d) are estimated directly from the hydrological data.

In the recent contributions by Starr, Peixoto, and Crisi (1965) and Peixoto and Crisi (1965), the water balance from the atmospheric water vapor flux has been computed for the whole Northern Hemisphere. The computation was, however, performed separately for the winter and summer seasons and for the whole year, in which case the term (a) can be considered negligible.

The purpose of the present paper is to describe and discuss the annual march of water vapor balance over the Hemisphere and over Asia, and the zonally summarized northward water vapor transfer for every month. To compute the water vapor balance (b), we calculated the terms (a), (c), and (d) from climatological data and obtained the term (b) as a residue of the water vapor balance equation.

The results represent an outline of water vapor balance over the Northern Hemisphere: The average precipitation over the Northern Hemisphere has a maximum of 104 mm cm² month^{-1} in summer and a minimum of 60 mm cm^{-2} month^{-1} in winter. The average evaporation over the Northern Hemisphere has a maximum of 94 mm cm²/month in winter and a minimum of 68 mm cm²/month in spring. The surplus evaporation over the precipitation has a seasonal variation with a maximum amount of 20 mm cm^{-2} month^{-1} in winter and a minimum of -30 mm cm^{-2} month^{-1} in summer. On the other hand, the time change of average water vapor storage in the Northern Hemispheric atmosphere has a seasonal variation with a maximum of 5 mm cm^{-2} month^{-1} in spring and a minimum of -5 mm cm^{-2} month^{-1} in autumn. These amounts show that on the average over the Northern Hemisphere the seasonal variation of water vapor storage in the atmosphere is not sufficiently smaller than that of the surplus evaporation over precipitation to be neglected. But finally the southward water vapor outflow across the equator, that is, evaporation minus precipitation minus water vapor storage change has almost the same seasonal variation as the surplus evaporation over the precipitation, and this means that the southward water vapor transfer is realized across the equator in winter and the northward transfer in summer.

2. FORMULAE FOR THE ATOMOSPHERIC WATER VAPOR BALANCE

If ω, the vertical wind component in the pressure coordinate ($\omega = dp/dt$), is assumed to become zero at the surface, the atmospheric water balance is expressed by

$$\frac{1}{g} \int_0^{p_s} \frac{\partial q}{\partial t} dp + \frac{1}{g} \int_0^{p_s} \nabla \cdot q\vec{V} dp = E - P, \tag{1}$$

where E denotes the evaporation from the surface, P the precipitation, g the gravitational acceleration, p_s the surface pressure, q the speicfic humidity, and $\vec{V}$ the horizontal wind vector.

If we write

$$W = \frac{1}{g} \int_0^{p_s} q\,dp, \qquad \vec{Q} = \frac{1}{g} \int_0^{p_s} q\vec{V} dp, \tag{2}$$

where W denotes the precipitable water and $\vec{Q}$ the integrated water vapor flux, Eq. (1) can be rewritten in the form

$$\frac{\partial W}{\partial t} + \nabla \cdot \vec{Q} = E - P. \tag{3}$$

The first term of Eq. (3) is the time change of precipitable water, the second the horizontal divergence of the vertically integrated water vapor flux, the third the evaporation and the fourth precipitation.

For our purposes, let us first evaluate $\partial W/\partial t$, E, and P from the climatic data and estimate $\nabla \cdot \vec{Q}$ from Eq. (3) for each month.

3. LOCAL TIME CHANGE OF MONTHLY PRECIPITABLE WATER $\partial \overline{W}/\partial t$

The dew-point temperature data are available in the *Humidity Atlas* published by the Gringorton *et al.* (1964). The atlas contains charts on the distribution of dewpoint at the surface, 850 mb, 700 mb, 500 mb, and 400 mb levels over the Northern Hemisphere for each midseason month: January, April, July, and October. The distributions are based on climatological data for 1958–1962 observed at 1,500 stations at the surface and 400 stations at the aerological levels. Dew-point amounts were read out at each intersection of 10 degree's latitude and longitude. Initially, the water vapor pressure in mb e was computed by following the Magnus formula substituting the dew-point amounts in Kelvin degrees into T_d:

$$\log_{10} e = -2{,}937.4/T_d - 4.9283 \log_{10} T_d + 23.5518. \tag{4}$$

Second, the specific humidity q was computed by the following equation substituting the pressure p in mb and the water vapor pressure e in mb:

$$q = 0.622 \, e/(p - 0.378 \, e). \tag{5}$$

Finally the precipitable water W in g was computed by the following formulae for vertical integration in the atmosphere, substituting the specific humidity at each level:

$$W = \tfrac{1}{2} \{166.2(q_s + q_{85}) + 152.8(q_{85} + q_{70}) + 203.7(q_{70} + q_{50}) +$$
$$101.9(q_{50} + q_{40}) + 407.0 \, q_{40}\}, \tag{6}$$

where q_s denotes the specific humidity at the surface and q_n the specific humidity at aerological levels of n cb. In formulating (6) we assumed that the specific humidity above the 400 mb level decreases linearly to zero at the 0 mb level. The precipitable water W for all months except January, April, July, and October was computed by interpolation with trigonometrical functions. To be precise, over the tropics the sine function of a half-year cycle with a maximum in April and October, and a minimum in January and July was applied in the interpolation. Over the other region, the sine function of a year cycle with a maximum in July and a minimum in January was applied. The distributions of climatic monthly mean precipitable water over the Northern Hemisphere have been given for all months by Kubota (1969).

The time change of monthly mean precipitable water was estimated by

$$\partial \overline{W}_m/\partial t = (\overline{W}_{m+1} - \overline{W}_{m-1})/2M, \tag{7}$$

where the suffix m denotes the order of the month, M the length of the month and the upper bar the monthly mean. The distributions were illustrated as given in the figures by Kubota (1969).

As a result, it was concluded that the time change of monthly mean precipitable water over the Northern Hemisphere is one order smaller as compared with the other terms in the water vapor budget Eq. (3). That is, the maximum value of the change of precipitable water is only 18.3 mm cm^{-2} month^{-1}, found at 30°N-80°E in the Ganges Basin in May, and the minimum -16.2 mm cm^{-2} month^{-1} at the same area in September.

The latitudinal difference of precipitable water is very large. The precipitable water over the equator is about three times that over 80°N even in summer, when the difference is relatively small, and is more than five times that over 80°N in winter. The time change of precipitable water over the tropics, on the other hand, is relatively small throughout the year. Naturally, the time change over high latitudes is also relatively small throughout the year. The time change over the mid-latitudes has the largest seasonal variation over the Hemisphere with the maximum in spring and the minimum in autumn.

The precipitable water over land is slightly smaller than over the ocean for almost all the latitudes troughout the year. On the other hand, it is interesting to note that the time change of precipitable water over land is larger than over the ocean and is smaller in autumn: that is, the time change over land has a much larger amplitude of seasonal variation than that over the ocean.

In May, the time change is at a maximum in the Hemisphere, except in the equatorial zone. Glancing at the distribution for May, two areas of high value are found over East Asia and over southern North America. The maximum amount over East Asia is about twice that over southern North America. The ridge of high time change over East Asia lies from the Ganges Basin to the Yellow Sea. In September, the time change in the hemisphere is at a minimum except for the equatorial zone. And it is interesting to note that almost the same features are found for the minimum in September as for the maixmum in May.

4. MONTHLY EVAPORATION $\bar{E}$

The data source of evaporation is the climatic monthly mean evaporation calculated by Budyko (1963), who used the following equation in calculating the evaporation over the sea surface:

$$\bar{E} = \frac{au}{L}(q^* - q), \tag{8}$$

where a is a proportional coefficient, independent of wind speed, 2.4×10^{-6}g cm^{-3}, L the latent heat, $597 + 0.6t$ (°C) in cal g^{-1}, u the wind speed in m sec^{-1}, q^* the specific humidity of saturated water vapor at the temperature of the vaporizing surface, and q the specific humidity of the air. Budyko used various techniques for calculating the evaporation over the land surface. We read out the monthly evaporation $\bar{E}$ at each intersection of 10 degree's latitude and longitude over the Northern Hemisphere from Budyko's Altas for each month. Table I shows the values obtained for the maximum and minimum evaporation over the Hemisphere.

Table I. Maximum and minimum monthly evaporation over the Northern Hemisphere in mm cm^{-2} month^{-1}.

	Jan.	Feb.	March	Apr.	May	June	July	Aug.	Sept.	Oct.	Nov.	Dec.
max.	319	252	219	175	155	205	188	188	202	235	275	319
min.	3	3	3	5	5	10	10	10	11	10	5	3

The minimum evaporation takes place over the Arctic Ocean throughout the year. The seasonal variation of the minimum evaporation accords with the seasonal variation of evaporation over the Arctic Ocean.

It is notable that the maximum evaporation takes place at 30°N–140°E near the Ogasawara Islands throughout the winter months (October–April). This proves that flourish evaporation rises over the Ogasawara high pressure area or the subtropical high pressure area.

In the summer months maximum evaporation takes place over the Pacific or the Indian Ocean south of 20°N. This shows that flourish evaporation does not rise over the subtropical high pressure area during the summer.

5. Monthly Precipitation $\bar{P}$

The main data source of climatic monthly precipitation used was WMO's Clino (WMO, 1962). Normals in Clino are averaged for 1951–1960, 1941–1960 or 1931–1960. Another data source was found in the climatography by Hatakeyama (1964) for China, Mongolia, Korea, Turkey, Indonesia, and Saudi Arabia. The period averaged for normal varies according to the stations.

Data at 814 stations over the Northern Hemisphere were plotted and isohyet lines were drawn. The values for every 10 degrees of latitude and longitude over the Northern Hemisphere were then read out for each month. The maximum and minimum values of $\bar{P}$ for all the months are shown in Table II.

Table II. The maximum and minimum values of $\bar{P}$ over the Northern Hemisphere in mm cm^{-2} month^{-1}.

	Jan.	Feb.	March	Apr.	May	June	July	Aug.	Sept.	Oct.	Nov.	Dec.
max.	400	305	420	480	450	450	500	540	420	460	407	450
min.	0	0	0	0	0	0	0	0	0	0	0	0

The minimum precipitation takes place in the Sahara Desert or in other deserts, and is zero throughout the year.

In the temperate zone, two narrow areas of heavy precipitation are found. One is the area from the Gulf of Alaska to the Coast Mountains in southeastern Alaska. The precipitation in the area is heavy throughout the year except from May to July, and is especially heavy in autumn. At the station Yakutat (59°31′ N, 139°40′ W) in southeastern Alaska, for example, there was an average precipitation of 3,348 mm/year and a seasonal maximum of 498 mm in October for 1931–1960. This settled heavy precipitation is thought to be caused by the surface circulation around the stationary low pressure over the northern Pacific Ocean and to be forced upstream on the slope side of the Coast Mountains. The other area is over the eastern Mediterranean Sea. The rain in the area is heavy only in winter. At the station Antalya on the south coast of Turkey, for example, the average precipitation for 21 years was 1,059 mm/year and the seasonal maximum was 267 mm in December.

In the tropics three broad areas with much heavier precipitation are found. The most important is Monsoon Asia. The second is the area from Upper Guinea to the Congo River Basin, and the third the area from the Caribbean Sea to northern South America. Heavy rain in these areas is related to the intertropical convergence zone, and it is evident that there is a seasonal north-south oscillation. Over Monsoon Asia, for example, the maximum rainfall area shifts to northern India in summer and to the tropics in the Southern Hemisphere.

6. Seasonal Variation of Zonally Averaged Evaporation and Precipitation

Zonally averaged evaporation and precipitation over the oceans, the continents and the entire circle of latitudes is shown in Fig. 1. Several features can be pointed out in the figure.

1) Except in the equatorial region, evaporation from the sea surface is more active in winter than in summer. Such an interesting seasonal variation must be caused by the fact

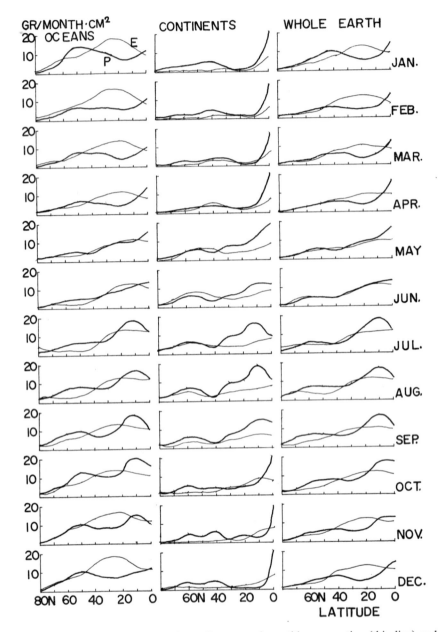

Fig. 1. Latitudinal distributions of zonally averaged monthly evaporation (thin line) and precipitation (thick line) over the oceans, continents and the whole latitudinal circle in $g\,cm^{-2}$ $month^{-1}$ for every month.

that the wind speed and the vertical thermal instability are both stronger in winter than in summer. Evaporation from the land surface is naturally more active in summer than in winter for all the latitudes, because the atmospheric boundary layer over the land is thermally more unstable in summer than in winter.

2) With respect to precipitation over the oceans, it is very clear that it has a peak at about 50°N throughout the year and that the peak is much higher in winter than in summer. Even over the land, we can find a peak at high latitudes, but the peak is not as clear, showing less seasonal variation. In the tropics, except in the equatorial zone, it is very noticeable that both over land and sea, the precipitation is much more active in summer and autumn than in winter and spring. This can be explained by the seasonal shifting of the intertropical convergence zones and the tropical high pressure zone. In the equatorial zone, the precipitation is much more active over the continents than over the oceans.

The average precipitation and evaporation for the whole Hemisphere is shown as a function of the month of the year in Fig. 2. The total precipitation or evaporation in the Northern Hemisphere equals these averages multiplied by the area of the Hemisphere. The total precipitation over the Northern Hemisphere is least in the late winter and greatest in summer. But the total evaporation is least in mid-spring and greatest in mid-winter. Evaporation exceeds precipitation from November to April and precipitation exceeds evaporation from May to October.

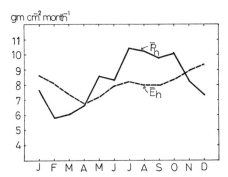

Fig. 2. Precipitation $\bar{P}_h$ (thick line) and evaporation $\bar{E}_h$ (thin line) averaged for the entire Northern Hemisphere as a function of the month of the year in g cm^{-2} year^{-1}.

The annual averages of mean hemispheric precipitation and evaporation are respectively 1,001.3 and 969.6 mm cm^{-2} year^{-1}. The excess of precipitation over evaporation is 31.7 mm cm^{-2} year^{-1}. If this amount is consistent over a long period, the water vapor corresponding to the total excess of precipitation, which equals 31.7 × (the area of the Northern Hemisphere) in mm year^{-1}, must be transferred to the Northern Hemisphere across the equator from the Southern Hemisphere by the atmospheric motion, and the water corresponding to the total excess must be transferred to the Southern Hemisphere across the equator by the sea currents and the river and subsoil discharge on the equatorial land masses.

7. Hemispheric Distributions of the Vertically Integrated Divergence of Water Vapor Flux

The vertically integrated divergence of monthly mean water vapor flux is computed by the following equation, substituting $\bar{E}$, $\bar{P}$, and $\partial \bar{W}/\partial t$:

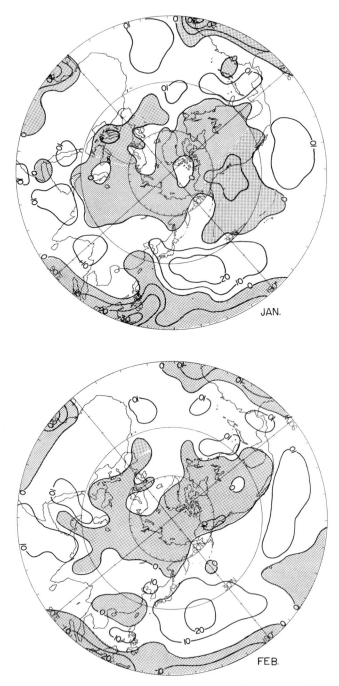

Fig. 3. Hemispheric distributions of the vertically integrated mean monthly divergence of water vapor flux $\overline{\nabla \cdot Q}$ (or mean monthly water vapor source) in g month^{-1} for every month.

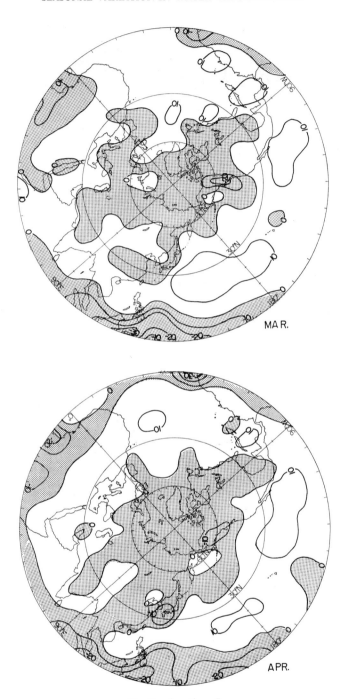

Fig. 3. (continued)

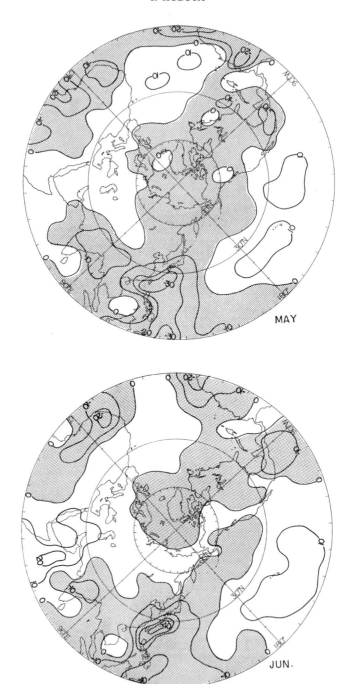

Fig. 3. (continued)

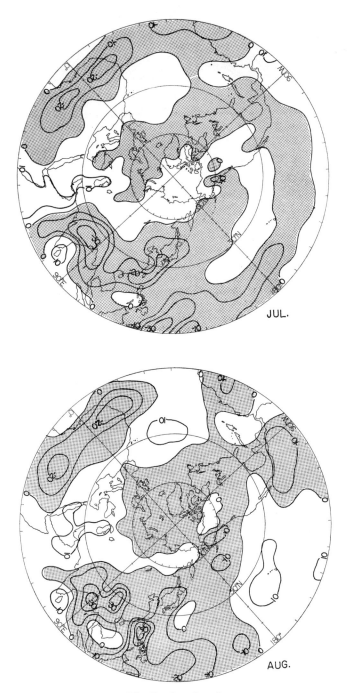

Fig. 3. (continued)

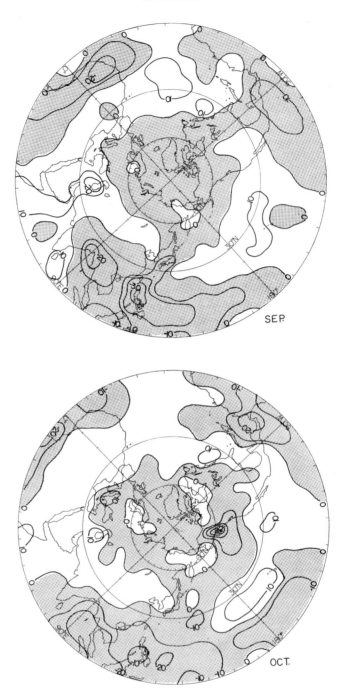

Fig. 3. (continued)

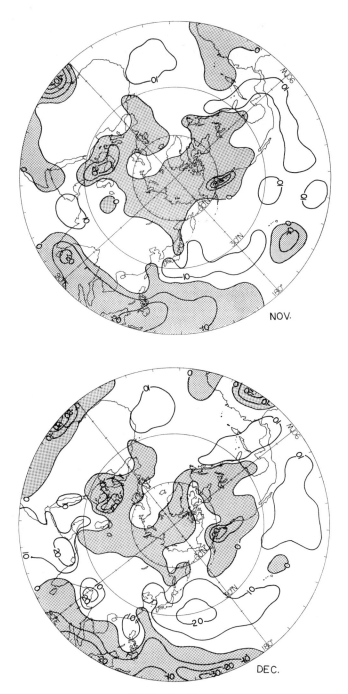

Fig. 3. (continued)

$$\overline{\nabla \cdot \vec{Q}} = \overline{E} - \overline{P} - \partial \overline{W}/\partial t. \tag{9}$$

$\overline{\nabla \cdot \vec{Q}}$ represents the water vapor source in the atmosphere. The hemispheric distributions of $\overline{\nabla \cdot \vec{Q}}$ for each month are shown in Fig. 3. The latitudinal distributions of zonally averaged $\overline{\nabla \cdot \vec{Q}}$ over the oceans, the continents and the whole latitudinal circle for each month are given in Fig. 4. We computed $\overline{\nabla \cdot \vec{Q}}$ for every 10 degrees of latitude and longitude over the Northern Hemisphere. The maximum and minimum values of $\overline{\nabla \cdot \vec{Q}}$ over the Northern Hemisphere are shown in Table III.

Table III. The maximum and minimum values of $\overline{\nabla \cdot \vec{Q}}$ over the Northern Hemisphere in mm cm^{-2} month^{-1}.

	Jan.	Feb.	March	Apr.	May	June	July	Aug.	Sept.	Oct.	Nov.	Dec.
max.	237	212	152	141	121	193	185	188	202	152	150	234
min.	−332	−235	−331	−363	−395	−417	−392	−452	−309	−443	−447	−335

Here let us compare the order of magnitude of each component in the water vapor budget Eq. (9). Remember that the range of variation of $\partial \overline{W}/\partial t$, the time change of precipitable water, is -16.2 mm cm^{-2} month^{-1} $< \partial \overline{W}/\partial t <$ 18.3 mm cm^{-2} month^{-1} over the Hemisphere and also throughout the year. Note Tables I to III, in which the range of variation of other components over the Hemisphere are shown for each month. The tables indicate that the time change of precipitable water is one order smaller than the other components in the water vapor balance equation. That is, $\overline{\nabla \cdot \vec{Q}}$, the vertically integrated divergence of water vapor flux, is approximately equal to $\overline{E} - \overline{P}$, the evaporation minus the precipitation for all the months.

The following points shown in Figs. 3 and 4 are noteworthy:

1)   As shown in Fig. 4, the values of $\overline{\nabla \cdot \vec{Q}}$ averaged zonally over the oceans and over the entire earth always show a divergence in the subtropical zone except in the summer months. The divergence area shifts north to 30°N in autumn and south to 17°N in spring. It is thought to correspond to the subtropical high pressure zone.

2)   As shown in Fig. 4, weak convergence zones, which may correspond to the polar frontal zones, are found near 50°N both over land and sea. A strong convergence zone, which corresponds to the intertropical convergence zone, is obviously found over the tropics and shifts to the northernmost latitude of 12°N in autumn. The strength of the convergence is larger over the continents than over the oceans.

3)   As shown quite strikingly in Fig. 4, convergence surpasses divergence over the continents, because precipitation is greater here than evaporation.

4)   As shown in Fig. 3, the systematic source is found over the subtropical western Pacific in the winter monsoon season, and the sink over Malaysia. On the other hand, in the summer monsoon season, the source is found over the Red Sea, the Caspian Sea, the Arabian Sea and the Persian Gulf, and the sink over the continental coastal region of Southeast Asia.

5)   As shown in Fig. 3, there are strong water vapor sinks over Southeast Asia, Middle West Africa and northern South America throughout the year.

6)   As shown in Fig. 3, two strong and narrow water vapor sink regions are found during the winter months, one over the southeastern coast of Alaska and the other over Europe. This implies that precipitation exceeds evaporation over these areas.

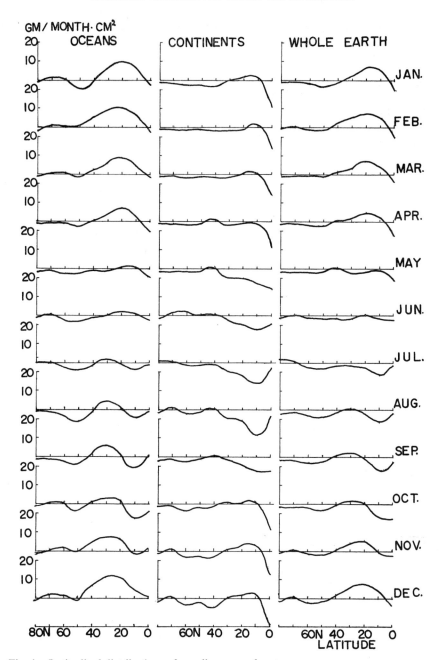

Fig. 4. Latitudinal distributions of zonally averaged water vapor source over oceans, continents and the whole latitudinal circle in g cm⁻² month⁻¹ for every month.

7)   It can also be seen that in the Arabian Sea strong water vapor sources are found throughout the year.

8. ZONALLY AND VERTICALLY INTEGRATED NORTHWARD WATER VAPOR FLUX

The evaporation minus precipitation averaged for the whole hemisphere $\bar{E}_h - \bar{P}_h$ was calculated from Fig. 2 and shown as a function of each month by a thin broken line in Fig. 5. The seasonal variation is maximal in summer and minimal in winter. The time change of precipitable water averaged for the whole hemisphere $\partial \bar{W}_h/\partial t$ is also plotted as a thin full line in Fig. 5. The amount is greatest (5 mm cm^{-2} month^{-1}) in May and least (-5 mm cm^{-2} month^{-1}) in October. The vertically integrated divergence of water vapor flux for the whole atmosphere $\overline{\nabla \cdot \vec{Q}_h}$ was calculated from the water vapor budget equation averaged for the whole Hemisphere:

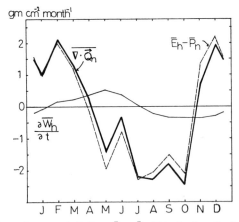

Fig. 5.   Evaporation minus precipitation $\bar{E}_h - \bar{P}_h$ (thin broken line), time change of precipitable water $\partial \bar{W}_h/\partial t$ (thin full line) and vertically integrated divergence of water vapor flux $\overline{\nabla \cdot \vec{Q}_h}$, averaged for the entire Northern Hemisphere in gm cm^{-2} year^{-1}.

$$\overline{\nabla \cdot \vec{Q}_h} = \bar{E}_h - \bar{P}_h - \frac{\partial \bar{W}_h}{\partial t}, \tag{10}$$

and shown by a thick full line in Fig. 5. In the water vapor budget equation averaged for the whole Hemisphere, it is found that the range of seasonal variation of $\partial \bar{W}_h/\partial t$ is about one fourth that of $\overline{\nabla \cdot \vec{Q}_h}$ and cannot be neglected in Eq. (10). The seasonal variation of $\overline{\nabla \cdot \vec{Q}_h}$ requires a southward water vapor transfer across the equator in the winter and a northward transfer in the summer. The vertically integrated divergence of monthly water vapor flux $\overline{\nabla \cdot \vec{Q}}$ is represented by the following equation:

$$\overline{\nabla \cdot \vec{Q}} = \frac{1}{R \cos \varphi} \left(\frac{\partial \overline{\{qu\}}}{\partial \lambda} + \frac{\partial \cos \varphi \overline{\{qv\}}}{\partial \varphi} \right), \tag{11}$$

where R denotes the earth's radius, φ the latitude, λ the longitude, u, v the wind components, upper bar the monthly mean, and $\{\ \}$ the vertical integration for the atmosphere. Integrating (11) with respect to $R \cos \varphi \, d\lambda$ cyclically along a latitudinal circle, we have the following equation:

$$\oint \overline{\nabla \cdot \vec{Q}} \ R \cos \varphi \, d\lambda = \frac{1}{R} \frac{\partial}{\partial \varphi} \oint \{\overline{qv}\} \ R \cos \varphi \, d\lambda$$

$$= \frac{1}{R} \frac{\partial}{\partial \varphi} \langle \{\overline{qv}\} \rangle, \tag{12}$$

where $\langle \ \rangle = \oint R \cos\varphi \, d\lambda$. Integrating (12) with respect to $R d\varphi$ from $\varphi = 90°N$ to $\varphi = \varphi$, we have

$$\int_{90°}^{\varphi} \left(\oint \overline{\nabla \cdot \vec{Q}} \ R \cos \varphi \, d\lambda \right) R d\varphi = \langle \{\overline{qv}\} \rangle_{\varphi=\varphi} - \langle \{\overline{qv}\} \rangle_{\varphi=90}. \tag{13}$$

As the zonally and vertically integrated northward water vapor flux $\langle \{\overline{qv}\} \rangle$ is zero at the Pole, Eq. (13) becomes

$$\langle \{\overline{qv}\} \rangle = \int_{90°}^{\varphi} \left(\oint \overline{\nabla \cdot \vec{Q}} \ R \cos \varphi \, d\lambda \right) R d\varphi. \tag{14}$$

Using Eq. (14) and substituting the climatic monthly values of $\overline{\nabla \cdot \vec{Q}}$, we computed the climatic monthly northward water vapor flux $\langle \{\overline{qv}\} \rangle$. The resulting latitudinal distributions $\langle \{\overline{qv}\} \rangle$ are given for each month in Fig. 6.

The following features are found in the distributions shown in Fig. 6: The northward water vapor transfer $\langle \{\overline{qv}\} \rangle$ in the tropics is reversed from winter to summer, i.e., it is southward in winter and northward in summer as estimated previously. There is also a weak peak of northward transfer of water vapor in the middle latitudes throughout the year.

9. Hemispheric Distribution of the Vertically Integrated Annual Divergence of Water Vapor Flux

Taking the annual average of the water vapor budget Eq. (9) as $\partial \overline{W}/\partial t = 0$, we have

$$\overline{\nabla \cdot \vec{Q}} = \bar{E} - \bar{P}, \tag{15}$$

where the thick bar represents the annual mean. The annual mean of the vertically integrated divergence of water vapor flux balances with the annual evaporation minus precipitation. The hemispheric distribution of $\overline{\nabla \cdot \vec{Q}}$ is shown in Fig. 7. For comparison's sake, the same distribution of $\overline{\nabla \cdot \vec{Q}}$ for 1950 computed aerologically by Starr and Peixoto (1958) is shown in Fig. 8. Comparing the two figures, the following similarities and differences are found:

1) $\overline{\nabla \cdot \vec{Q}}$ or $\bar{E} - \bar{P}$ averaged over the oceans is plus and over land is minus. In our computation, $\overline{\nabla \cdot \vec{Q}}$ or $\bar{E} - \bar{P}$ averaged over the oceans is about 141 mm cm^{-2} year^{-1} and over land, about -345 mm cm^{-2} year^{-1}. The latter amount multiplied by the total area of land in the Northern Hemisphere must be outflowed to the oceans owing to river and subsoil discharge. As the annual water vapor transfer between the Northern and Southern Hemispers can be neglected, as estimated in section 6, the latter amount multiplied by the total area of land must equal the first amount multiplied by the total area of ocean in the Northern Hemisphere.

2) Common features are the wide and intensive divergences over the Arabian Sea, the

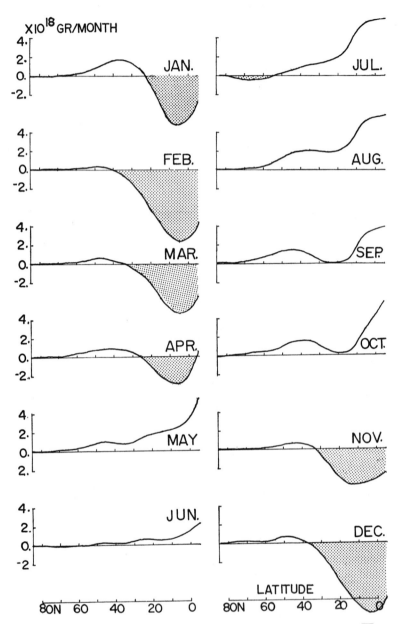

Fig. 6. Latitudinal distributions of monthly northward water vapor flux $\langle \overline{\{qv\}} \rangle$ in 10^{18} g month^{-1} for every month.

Fig. 7. Hemispheric distribution of vertically integrated annual divergence of water vapor flux $\overline{\nabla \cdot \vec{Q}}$ $(=\bar{E}-\bar{P})$ in g year^{-1} computed in this work.

Fig. 8. Hemispheric distribution of vertically integrated annual divergence of water vapor flux $\overline{\nabla \cdot \vec{Q}}$ in g year^{-1} computed for 1950 aerologically by Starr and Peixoto (1958).

central and eastern parts of the northern Pacific and northern Atlantic Oceans, equatorial Africa and near the equator in South America.

3)   As to differences, in the distribution obtained in this work (Fig. 7), there is a wide and intensive divergence over Southwest Asia. This divergence region is formed by many strong divergence areas over the seas in Southwest Asia—the Arabian Sea, the Persian Gulf, the Red Sea, and the Caspian Sea. A strong divergence was found over Mexico in the distribution in Fig. 8, but it was not found in Fig. 7.

4)   In the distribution in Fig. 7, an extremely intensive convergence center appears over southern Alaska, due to the fact that the annual precipitation is higher than the evaporation there. This intensive convergence was not found in Fig. 8, perhaps because of the type of data used: there is little aerological data for the Northeast Pacific Ocean.

5)   Over Borneo and Celebes, the convergence in Fig. 7 is much more intensive than that in Fig. 8. One reason for this may be that the precipitation over the oceans estimated from that at the stations on the islands exceeds the actual amount.

6)   It is interesting to note that there are weak divergences over the Sahara Desert and Kazakh in the U.S.S.R. in both figures, and over Mongolia in Fig. 7. If the divergences are appreciable, river and subsoil discharges must carry the water into the regions.

7)   In Fig. 7, a convergence exceeding 15 g cm^{-2} year^{-1} is found over the area centered in the Bashi Channel, but in Fig. 8, the corresponding convergence is not there but over the area east of the Ogasawara Islands.

8)   In Fig. 8, there is a strong divergence over the area from the Japan Sea to the East China Sea, but no corresponding divergence is found in Fig. 7.

Acknowledgement

The author wishes to thank Dr. M. M. Yoshino for his support in the publication of this work and his many suggestions after reading this paper.

References

Adem, J. 1962: On the theory of the general circulation of the atmosphere. *Tellus* **14** 102–115.

Adem, J. 1965: On the normal thermal state of the troposphere-ocean-continent system on the Northern Hemisphere. *Geofisica International* **4**(1) 3–32.

Adem, J. 1967: Parameterization of atmospheric humidity using cloudiness and temperature. *Mon. Wea. Rev.* **95** 83–88.

Arakawa, A. 1958: Recent theories of general circulation. *Kisho Kenkyu Note* **9**(4) 1–134 (in Japanese).

Atlas, D. 1965: Model atmospheres for precipitation. *In* Handbook of Geophysics and Space Environments (Shea L. Valley, ed.) **5** 6–11, AFCRA.

Berliand, M. E., Berliand, T. G. 1952: Determination of the effective outgoing radiation of the earth, taking into account the effect of cloudiness. *Izv. Akad. Nauk SSSR, Ser. Geofiz.* (1) 64–78.

Budyko, M. I. 1956: Teplovoĭ balans zemnoĭ poverkhnosti. Gidrometeorologiyaeskoe izdatelvstvo, Leningrad.

Budyko, M. I. 1963: Atlas teplovogo balansa zemnogo shara. Mezhduvedomstvennyĭ Geofizicheskiĭ Komitet Pri prezidium, Akademiĭ Nauk SSSR, Glavnaya Geofizicheskaya Observatoriya imenniĭ A. E. Voeĭkova Resultaty, Moskva.

Gringorten, I. I., Salmela, H. A., Solomon, I., Sharp, J. 1966: Atmospheric Humidity Atlas—Northern Hemisphere. AFCRL-66-621.

Hatakeyama, H. (ed.) 1964: Climate in Asia. Kokin Syoin, 577p.

Jacobs, W. C. 1950: The distribution and some effects of the seasonal quantities E-P (evaporation minus precipitation) over the North Atlantic and North Pacific. *Arch. Met. Geoph. Biokl.* Ser. A **1**(1) 1–16.

Katayama, A. 1962: Heat budget on the earth. *Kisho Kenkyu Note* **13** 101–169.

Katayama, A. 1966: On the radiation budget of the troposphere over the Northern Hemisphere (1). *J. Met. Soc. Japan* **44** 381–401.

Katayama, A. 1967: On the radiation budget of the troposphere over the Northern Hemisphere (2), (3). *J. Met. Soc. Japan* **45** 1–39.

Kubota, I. 1967: Distribution of mean monthly precipitable water in the Northern Hemisphere and its time change. *Climatological Notes, Dept. Geography, Hosei University* (2) 1–28.

Kurosaki, A. 1967: Problem of the boundary layer on general circulation. *Grosswetter* **5**(3) 45–50 (in Japanese).

Manabe, S., Möller, F. 1961: On the radiative equilibrium and heat balance of the atmosphere. *Mon. Wea. Rev.* **89** 503–532.

Manabe, S., Strickler, R. F. 1964: Thermal equilibrium of the atmosphere with a convective adjustment. *J. Atmos. Sci.* **21** 361–385.

Manabe, S., Smagorinsky, J., Strickler, R. F. 1965: Simulated climatology of general circulation with a hydrologic cycle. *Mon. Wea. Rev.* **93** 769–798.

Manabe, S., Wetherald, R. T. 1967: Thermal equilibrium of the atmosphere with a given distribution of relative humidity. *J. Atmos. Sci.* **24** 241–259.

Peixoto, J. P. 1960: On the global water vapour balance and the hydrological cycle. *In* Tropical Meteorology in Africa 232–243.

Peixoto, J. P. 1965: On the role of water vapor in the energetics of the general circulation of the atmosphere. *Portugalia Physica* **4** 135–170.

Peixoto, J. P., Crisi, A. R. 1965: Hemispheric humdity conditions during the IGY. *Sci. Rpt. No. 6, Planetary Circulations Project.* M.I.T.

Starr, V. P., Peixoto, J. P., Grisi, A. R. 1965: Hemispheric water balance for the IGY. *Tellus* **17** 463–472.

Starr, V. P., Peixoto, J. P. 1958: On the global balance of the water vapor and hydrology of deserts. *Tellus* **10** 188–194.

Starr, V. P., Peixoto, J. P. 1960: On the zonal flux of water vapor in the Northern Hemisphere. *Geof. Pura e. Appl.* **47** 199–203.

Starr, V. P., Peixoto, J. P. 1964: The hemispheric eddy flux of water vapor and implications for the mechanics of the general circulation. *Arch. Met. Geoph. Biokl.* Ser. A **14** 111–130.

Starr, V. P., Peixoto, J. P., Livadas, G. C. 1958: On the meridional flux of water vapor in the Northern Hemisphere. *Geof. Pura e Appl.* **39** 174–185.

Starr, V. P., White, R. M. 1954: Balance requirements of the general circulation. *Geophys. Res. Directorate AFCRC.*

Starr, V. P., White, R. M. 1955: Direct Measurement of the hemispheric poleward flux of water vapor. *J. Mar. Res.* **14** 217–225.

Shono, S. 1954: Kisho Rikigaku Josetsu. 425p. (in Japanese).

Takahashi, K. et al. 1957: Heat budget of the atmosphere. *Kisho Kenkyu Note* 5.

U. S. Navy 1963: Marine climatic atlas of the world (6), Arctic Ocean. NAVWEPS 50–lc–533.

WMO/OMM-NO.117.TP.52, 1962: Climatological normals (Clino) for climate and climate ship stations for the period 1931–1960.

Yoshino, M. M. 1969: Climatological studies on the polar frontal zones and the intertropical convergence zones over South, Southeast, and East Asia. *Climatological Notes, Dept. Geography, Hosei University* (1) 1–71.

Zubenok, L. I., Strokina, L. A. 1963: Evaporation from the surface of the globe. *Main Geoph. Obs.* (Tr. GGO) **139** 93–107 (translation in Soviet Hydrology **6** 597–611, 1963).

SEASONAL CHANGE OF MOISTURE DISTRIBUTION IN MONSOON ASIA

Yoshiharu Neyama

Abstract: An analysis was made of the seasonal change in the saturation rate distribution and airflow pattern on the monthly mean 850 mb level in 1963 and 1966. It was found that a departure from the normal precipitation during the rainy season in Monsoon Asia shows a symmetrical distribution, and a new SW monsoon system corresponding to a heavy rain in eastern Monsoon Asia was observed. The appearance of the SW monsoon was caused by a northward shifting of the region with a strong ascending current following the strengthening of the easterly jet stream, produced by the strengthening and change of position of the zonal axis of the South Asian anticyclone. One important finding was that, in spite of the different precipitation distribution in the respective years in the western and central parts of Monsoon Asia, the moisture distribution and the process of occurrence of the SW monsoon showed no remarkable differences in either year.

1. INTRODUCTION

When the seasonal change of moisture distribution in Monsoon Asia is studied, the SW monsoon and the convergence zones in each air mass should be considered in the lower layer of the troposphere. Research of this type has been current since the turn of the century: but the principal studies date from the development of successful sounding method. Wagner (1931) and Ramanathan *et al.* (1939) gave the height of the SW monsoon in India as 4–6 km. Dao and Chen (1957) noted that the SW monsoon had a height of 3–8 km, belonged to the equatorial westerlies, and was separated from the subtropical westerlies by the Himalayas and from the easterlies. The occurrence of the SW monsoon was attributed to the northward transition of the westerly jet stream over the Himalayas (Yin, 1949). The new jet stream appears north of the Himalayas when the jet stream south of it disappears due to the dissolution of the south-north temperature difference caused by the Himalayas as the heat source in the upper layer (Flohn, 1957). Flohn (1957) and staff members of the Academia Sinica in Peking (1958) noted that the SW monsoon appeared with the advance of the monsoon trough toward the north in the lower troposphere and with the jumping of the westerlies over the Tibetan Plateau in the upper analysis of general circulation over East Asia. Yin (1949) suggested that the onset of the monsoon was caused by the westward march of the trough on the mean flow pattern at the 500 mb level. He supported the theory that the SW monsoon originates in the Northern Hemisphere, and not in the Southern Hemisphere (Depperman, 1940; Riehl, 1948; Thompson, 1951). The date of onset of the SW monsoon was calculated as May 25 at Ceylon by Thompson (1951), May 29 at Travancore-cochin and June 8 at Kolaba by Ramdas (1949), and April 28 at West Java by Schmidt and Vecht

(1952). The geographical locations of the Intertropical Convergence Zone in the figures shown by Sawyer (1952), Thompson (1951), Riehl (1954), Flohn (1957) etc. differ quite a bit from each other in consequence of the different definitions of the convergence zone as a boundary among the tropical, equatorial and polar air-masses (Yoshino, 1969). Ramanathan (1954) pointed out that most of the rainfall in India, Burma and Pakistan was caused by a turbulence in the upper layer during the monsoon month. The variation of the moist tongue in all areas of Monsoon Asia was analyzed by Thompson (1951) and the change of transfer of water vapor during the rainy season was discussed by Murakami (1959). The rainy season in East Asia was divided into four stages by considering the position of the frontal zones at sea level and the characteristics of the westerly wind at the 500 mb level by Yoshino (1965, 1966). As mentioned above, researches on various aspects of the SW monsoon have a long history, but no investigations have been made up to now of the airflow pattern and the simultaneous moisture distribution.

In this paper, the seasonal change of moisture distribution was studied in connection with the airflow pattern from the point of view of the vertical structure in Monsoon Asia. Therefore, the original convergence zone and monsoon corresponding to the moisture distribution were considered irrespective of the so-called NITC. The differences in process and situation of these in the years in which the total precipitation was very heavy and those in which it was slight were studied in terms of the vertical structure. The data used were the monthly climatic data for the world and the Northern Hemisphere daily data tabulations issued by the U.S.A., the aerological data of Japan, and the stratospheric charts issued by the Free University of Berlin.

2. SEASONAL CHANGE OF PRECIPITATION IN MONSOON ASIA

Since a distinct rainy season appears during the warm season in Monsoon Asia, the normal precipitation in each month is shown in Table I for comparative purposes. Aden, Teheran, Karachi, Bombay, Trivandrum, Calcutta, Bangkok, Singapore, Taipei, Hiroshima, and Tokyo were selected as representative sites in the area. It can be seen from Table I that the date of onset of the monsoon or the date on which the total monthly precipitation is noticeably greater than that of the previous month differs from place to place. In those

Table I. Normal monthly total precipitation (mm) at each station in Monsoon Asia. Normals are a 30-year average from 1931 to 1960.

Station \ Month	April	May	June	July	Aug.	Sept.
Aden	0	1	0	3	2	7
Teheran	31	14	2	1	1	1
Karachi	2	0	7	96	50	15
Bombay	3	16	520	709	419	297
Trivandrum	122	249	331	211	164	123
Calcutta	43	121	259	301	306	290
Bangkok	89	166	171	178	191	306
Singapore	160	101	127	183	230	102
Manira	24	110	236	253	480	271
Hong Kong	133	332	479	286	415	364
Taipei	182	205	322	269	266	189
Hiroshima	158	154	249	250	116	216
Tokyo	135	131	182	146	147	217

areas east of Bombay, i.e., Trivandrum, Bangkok, and Hong Kong, the rainy season starts in May; in other areas it starts in June. The years of 1963 and 1966 were taken as representative years with respect to the relationship between the moisture distribution and the airflow pattern at the 850 mb level. These years show opposite conditions with reference to the anomaly of total precipitation in June and July in each location as shown in Table II. The anomaly at Calcutta, Bangkok, Singapore, and Manila indicated in the right column of the table is positive in 1963 and negative in 1966, while the reverse is true in western India and in the areas east of Hong Kong. The amount of precipitation in the Baiu period in Japan is less than normal in 1963, and more than normal in 1966. Because the distribution of the departure from normal precipitation in other years was very complex for the various sites, the years of 1963 and 1966 were studied in detail in this paper.

Table II. Departure from the normal precipitation for the months at each station in Monsoon Asia in 1963 and 1966.

Station	Month / Year	April	May	June	July	Aug.	Sept.	J.+J.
Aden	1963	+2	−1	0	+5	0	−7	
	66	0	−1	0	−3	−1	−7	
Teheran	1963	−6	+24	−1	−1	+11	0	
	66	−20	+2	−1	0	−1	−1	
Karachi	1963	+1	0	−7	−96	−38	−15	
	66	−2	0	−7	−21	−50	−15	
Bombay	1963	−3	+16	−150	+88	+585	+33	−62
	66	−3	−14	−178	+175	−288	−114	−3
Trivandrum	1963	+6	+68	−162	+51	+43	+68	−111
	66	−69	−220	−62	−49	−108	+376	−111
Calcutta	1963	+34	0	+39	−10	−92	+105	+29
	66	−35	−99	+144	−169	−31	−87	−25
Bangkok	1963	−29	×	−51	−48	+129	+84	+181
	66	−17	+215	+44	+137	−34	−49	−99
Singapore	1963	−110	+29	+33	+7	−120	−72	+40
	66	+4	+103	−78	−15	−96	−38	−93
Manira	1963	−23	−107	+135	+65	−191	+409	+200
	66	−23	+330	−136	−11	−280	+404	−147
Hong Kong	1963	−119	−326	−274	+38	−233	−281	−236
	66	+204	−192	+484	+188	−197	−340	+672
Taipei	1963	−136	−96	−91	−90	−115	+446	−181
	66	+16	−139	+274	−155	+25	+526	+119
Hiroshima	1963	−9	+246	+5	−89	+184	+48	−84
	66	−12	+21	+78	−76	−65	+229	+2
Tokyo	1963	−59	+5	+66	−80	+235	−73	−14
	66	−1	+65	+328	+15	−92	−20	+343

3. SEASONAL CHANGE OF SATURATION RATE AND AIRFLOW PATTERN AT THE 850 mb LEVEL

Usually, the 850 mb level is used to detect the front, to examine its strength and weakness, and to determine the distribution of moisture by a map analysis. Therefore, to study the seasonal change in moisture distribution in Monsoon Asia, isolines representing a saturation rate, or difference between temperature and dew-point temperature at 850 mb, of 4°C or less are shown for April through September in 1963 in Fig. 1 and for 1966 in Fig. 2, in addition to the airflow pattern. A saturation rate of 4°C is used as a humidity index to express the distribution of water vapor in the NAWAC Manual (Synoptic Meteorological Center)

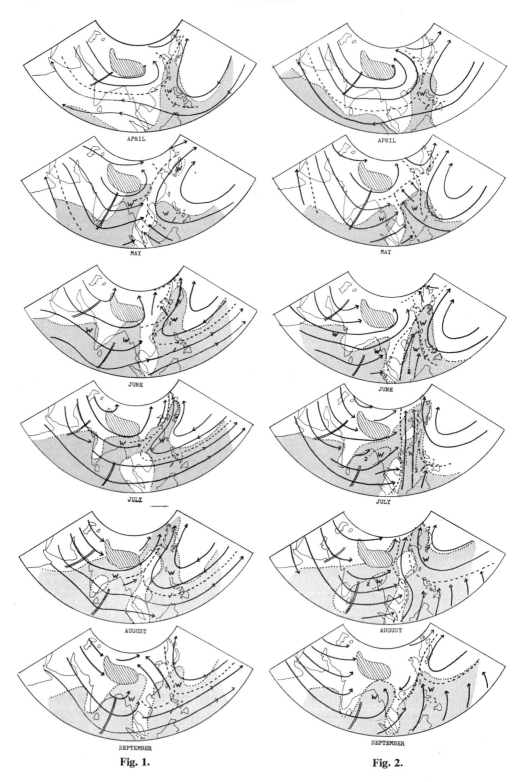

APRIL   APRIL

MAY   MAY

JUNE   JUNE

JULY   JULY

AUGUST   AUGUST

SEPTEMBER   SEPTEMBER

Fig. 1.   Fig. 2.

issued by the U.S.A. Figure 3 indicates the network of observations used in Figs. 1 and 2. The characteristics of seasonal change can be observed as follows:

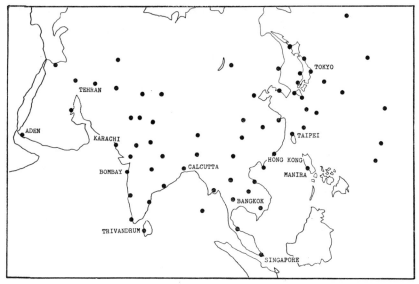

Fig. 3. Network of aerological stations.

a. *April*

First of all, the easterlies from the western Pacific toward the east coast of Africa along the low latitudes and the westerlies surrounding the Tibetan Plateau southeast of the Asian Continent occur in April. The subtropical SE monsoon flowing along the margin of the Pacific anticyclone and the flow changing from SW to SE around the Tibetan Plateau are seen in April. There are clear boundaries (convergence zones) of different wind directions between them.

The wet area, in which the saturation rate is 4°C or less, spreads from the South China Sea toward the East China Sea along the convergence zones. The trough extending toward the northwest of India from west of the Tibetan Plateau is located in the westerly flows at high latitudes. The remarkable difference between both years is the position of the boundary between the westerlies and the easterlies located to the south off the continent: the boundary in 1963 runs considerably north, extending from middle India toward China. There is not, however, a great difference between both years with respect to moisture distribution.

b. *May*

In May, the boundary mentioned above disappears, and the tropical westerlies blow in Monsoon Asia. The common phenomenon in both years is that the trough in the westerlies moves eastward, occupying a position from south of the Tibetan Plateau to west of the Bay

---

Fig. 1.   Monthly mean airflow and temperature patterns at the 850 mb level in 1963. The solid line indicates the airflow, the double line the trough, the dashed line the convergence zone or front, and the dotted line 4°C of temperature subtracted by dewpoint temperature. The shaded area shows wet regions.

Fig. 2.   Same as Fig. 1, but for 1966.

of Bengal, roughly along 82°E. The SW monsoon appears in the Bay of Bengal, the moist air mass is also present there, and the rainy season sets in over Burma at the same time. Part of the equatorial westerlies appears south of the Indochina Peninsula, the SE monsoon is intensified in South China, and invades Central China. The rainy season in Hong Kong begins as a result of the inflow of wet air mass accompanied by this monsoon circulation. But this SE monsoon is the flow which originated as the subtropical SE monsoon surrounding the Pacific anticyclone and then branched off in the South China Sea, coming over the Indochina Peninsula to southern China. The flows from the Pacific anticyclone were stronger in 1963 than in 1966, but the branched SE monsoon appeared only in the neighborhood of the Gulf of Tonkin in May of 1963. The area of the subtropical SE monsoon was relatively narrow in 1966, extending over the South China Sea to China. The equatorial westerlies seen in the vicinity of the Indochina Peninsula in 1966 were stronger than in 1963.

c. *June*

The most notable points accompanying the seasonal transition from May to June are *i*) the occurrence of the SW monsoon toward India east of the trough mentioned above and *ii*) the setting-in of the rainy season according to the expansion of the moist area toward the west coast of India. These phenomena are similar in both years. The trough located west of the Bay of Bengal in May moved more or less eastward to the middle of the Bay, and the SW monsoon blew continuously over Burma and the Malay Peninsula.

A remarkable difference in the patterns at the 850 mb level in both years is recognized in the regions from the Indochina Peninsula to the East China Sea through the east coast of China and at low latitudes in the western Pacific. In other words, the equatorial westerlies which appeared south of the Indochina Peninsula in May, 1963 came to blow eastward along the tropics south of 10°N in June. As a result the intertropical convergence zone extended eastward from north of the Indochina Peninsula through the South China Sea, and situated itself in the western Pacific parallel with its latitude in 1963.

On the other hand, in 1966, the equatorial SW monsoon, which differs from the SW monsoon coming from the Indian Ocean, appeared in the South China Sea and the Borneo Sea, west of the subtropical SE monsoon area. In the region from South China to the Malay Peninsula, the boundary of the two monsoon flows mentioned above can be found. This boundary can be recognized as one part of the convergence zone. The other part of the convergence zone jumped far northward, and came to extend from the vicinity of Japan via Manchuria to northern China. The wet tongue extending toward Japan became stronger. The conditions of moisture distribution in India and Burma in June are similar in both years. The moisture distribution over East Asia is also quite similar in both years, but the conditions are different: the moisture is formed by the tropical air mass following the subtropical SE monsoon in 1963, but by an equatorial air mass produced with the SW monsoon in 1966. Therefore, Hong Kong, Taipei, and Tokyo received a heavy rainfall as indicated by the positive departure of $+300 \sim +400$ mm from the June normal (cf. Table II). Presumably this anomalous precipitation could have been foretold by the occurrence of the SE monsoon branching off from the subtropical SE monsoon in the flow pattern in May.

d. *July*

The trough in the Arabian Sea which had extended into the equatorial westerlies in June remained almost stationary and the SW monsoon developed in western India. Moreover, the trough in the Indian Ocean remained constant and forced the SW monsoon in Burma to the mountainous regions. The changes from the previous month are, first, the strengthening of the SW monsoon in the whole region of Monsoon Asia, from India to Manchuria going round the Tibetan Plateau. The expansion of the moist area from eastern China to the Japan Sea is striking in 1963, but the location of the boundary between the subtropical SE monsoon and the SW monsoon is very similar in June and July. The fact that the equatorial westerlies blew along the low latitudes in the western Pacific is again similar to the previous month. No invasion of the equatorial air mass, brought by the equatorial SW monsoon, is seen in the eastern part of Asia in this month, either. The SW monsoon, reaching northern China, goes around the Tibetan Plateau also, but it is relatively weak in 1966. The southerly monsoon, which carried the equatorial air mass in June, became stronger in 1966. It flew into eastern Siberia from the South China Sea changing to a south wind. The intertropical convergence zone, the boundary between the subtropical SE monsoon and the SW monsoon, reached the northeastern part of the Tibetan Plateau passing through eastern Siberia in 1966. The convergence of the subtropical SE monsoon, the equatorial SW monsoon and the tropical SW monsoon was remarkably apparent in the region from southeastern part of Siberia to Manchuria, and it became wetter and expanded in area.

e. *August*

In August, the trough extending into the Arabian Sea became weaker, the trough in the Bay of Bengal advanced westward, and the flow system going around the Tibetan Plateau was still distinctly developed in both years. In southern India, the air current changed from WNW to NW, but the wet area spread to the north of India. The area is larger in August than in July. The roundabout flow in 1966 increased in intensity compared with that in 1963, so the moisture regime covering eastern Mongolia and Manchuria appeared clearer in 1966 than in 1963. On the other hand, the air flow system over East Asia in 1963 is almost similar to that in July, but the expanse of wet area covering the Yellow Sea, Manchuria, and southeastern Siberia differed slightly from that in July. The equatorial SW monsoon appeared over a wide area between the western Pacific and the Philippines in 1966, and consequently the northern boundary of wet area reached the latitude of 20°N in the western Pacific.

f. *September*

The most notable seasonal change from August to September in both years is that the large air flow rounding the Tibetan Plateau was divided at the western and eastern parts of the Plateau, forming separate flows.

In 1963, the trough in the Bay of Bengal was remarkable, a NW wind prevailed over India, and there was an airflow southeast of the Tibetan Plateau, but the airflow over Pakistan and India changed into a southerly wind in 1966. The common fact was the spreading of the moisture area from the Bay of Bengal, southeast of the Tibetan Plateau. Over the area from China to the South China Sea, the airflow became southeasterly in 1963, but northeasterly in 1966. Accordingly, the boundary between the subtropical SE monsoon and equatorial SW monsoon in 1966 shifted southward until it ranged north of the Indochina Peninsula from the East China Sea through the north of the South China Sea. The equatorial SW monsoon and the wet area over the western Pacific in both years did not change in August.

g. *Summary*

The monthly mean seasonal changes in the moisture distribution and the airflow system at the 850 mb level are given in detail above. The important facts can be summarized as follows:

1)   In the eastern part of Monsoon Asia, the SW airflow consists only of the subtropical SW monsoon blowing over India, Burma etc. in some years, and of both the subtropical SW monsoon and the equatorial SW monsoon (nearly a southerly wind) in other years. The fact that the equatorial westerly wind is sometimes southwesterly and sometimes not can be explained by considering that the flows transit from the Southern Hemisphere to the Northern Hemisphere, or that the equatorial air mass in the Northern Hemisphere is the cause. It is interesting to note that the SW monsoon over the Indian Ocean and the Bay of Bengal is always southwest, and the airflow systems over the western Pacific and East Asia differ from year to year.

2)   The moisture distribution in the rainy season over the Indian Ocean and the Bay of Bengal are almost similar, even if the rainfall distribution differs for each year. The process of the seasonal change and content of the moisture distribution, however, differ from year to year in the regions over East Asia; the Indochina Peninsula, South China, the East China Sea, and Japan. It is suggested that these are due to the important role played by the Tibetan Plateau in converging the current toward the Pacific Ocean, and in producing the prevalence of zonal current over South Asia.

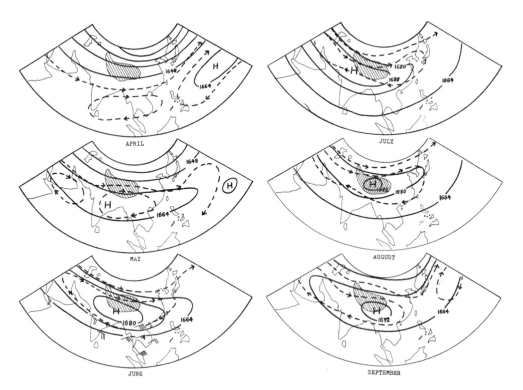

Fig. 4.   Monthly mean 100 mb contour and airflow patterns in 1966. The solid line indicates the contour, the dashed line the stream line; the wind speed is given in m/s.

4. VERTICAL STRUCTURE OF THE MONSOON

Many authorities mention that the appearance of the SW monsoon is associated with the transition of the westerly jet stream north of the Tibetan Plateau, and that the easterly jet stream in the lower stratosphere follows it simultaneously. The occurrence of the SW monsoon in the middle and western part of Monsoon Asia can be explained by the occurrence and movement of the trough shown by the wind shear at the 850 mb level. The airflow system contributing to the moisture distribution in eastern Monsoon Asia showed a remarkable difference between the years of 1963 and 1966, as noted above.

Starting with the anticyclone in the upper troposphere over South Asia, the monthly change of the anticyclone centered over the Tibetan Plateau at the 100 mb level in 1966 (Fig. 4) revealed that weak westerlies prevail over Monsoon Asia in April, and an anticyclonic circulation area is seen over the Bay of Bengal. In May, the anticyclone centered over the Indian Peninsula covers the whole area of Monsoon Asia, and the zonal axis of the circulation system reaches 20°N. The center of the anticyclone shifts northward in June, reaching the southern part of the Tibetan Plateau with a height of 16,800 gpm. The circulation system reaches 30°N, and the easterly jet stream flows stronger than 55 m/s over the Indochina Peninsula. In July, the anticyclone develops and advances northward, reaching a peak with the height of its center at 16,900 gpm. It is located over the Tibetan Plateau in August. In September, the anticyclonic center headed southeast of the Tibetan Plateau again, and the anticyclone weakened rapidly.

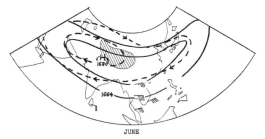

JUNE

Fig. 5. Monthly mean 100 mb contour and airflow patterns in June, 1963. Legend as in Fig. 4.

The conditions in June, 1963 are given in Fig. 5. The height of the anticyclone located south of the Tibetan Plateau is 16,800 gpm and its area is limited: i.e., the anticyclone is weaker in 1963 than in 1966. The velocity of the easterly jet stream over the Indochina Peninsula is about 45 m/s, which is weaker than in 1966. The anticyclonic circulation over eastern Monsoon Asia is stronger in 1966 than in 1963, and the east-west axis of the anticyclone in 1963 is much further northward than in 1966. The vertical profiles of vector wind are shown in Fig. 6 (at Clark) and Fig. 7 (at Naze) in order to examine the vertical construction of the atmosphere over these areas. In these figures, it is seen that the easterly wind in the whole troposphere at Clark is stronger in 1966 than in 1963, and the westerly wind appears in the layer below the mid-troposphere in 1963. Consequently, it can be said that the circulation system at low latitudes appeared already at Clark in June, 1966. The middle latitude westerly wind at Naze is much stronger in 1966 than in 1963 as the subtropical westerly jet stream moves southward and the vertical wind shear becomes very large over Naze. It is apparent from the wind direction at the 100 mb level as shown in Figs. 4 and 5 that the east-west axis of the South Asian anticyclone is situated south of Naze in 1966 and north of Naze in 1963. The general characteristics of the tropical easterly jet stream suggests

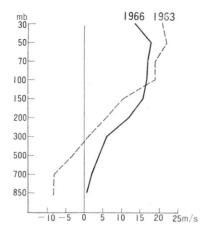

Fig. 6. Vertical distribution of monthly mean wind speed at Clark in June. Plus means easterly component.

Fig. 7. Same as Fig. 6, but at Naze.

(Flohn, 1964) that the strengthening of the subtropical westerly jet stream in the troposphere and the tropical easterly jet stream in the upper troposphere over East Asia in 1966 produces a large divergence in the higher layer and convergence in the lower layer in the vicinity of the entrance of the easterly jet stream, and that the ascending current at the right side of the entrance of the easterly jet stream becomes stronger. One finding that supports this suggestion is the fact that the layer below the 300 mb level at Hong Kong and Taipei indicated in Figs. 8 and 9 is wetter in 1966 than in 1963. Either the equatorial westerlies at the 850 mb level shifted remarkably northward as mentioned above or the strong easterly jet stream and the anticyclonic circulation system shifted far south in 1966. In 1963, on the other hand, the equatorial westerly wind moved toward the western Pacific at low latitudes. It is thought that the strengthening of the South Asian anticyclone in 1966 is due to the strong effect of the Tibetan Plateau as a heat source.

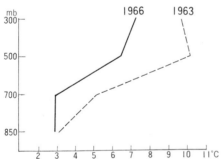
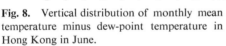
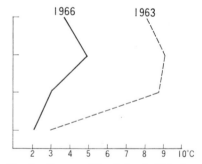

Fig. 8. Vertical distribution of monthly mean temperature minus dew-point temperature in Hong Kong in June.

Fig. 9. Same as Fig. 8, but in Taipei.

5. CONCLUSIONS

The monthly variation in moisture distribution during the warm season in Monsoon Asia was studied using the saturation rate, or the difference between air temperature and dew-point temperature, for 1963 and 1966, because the distribution of monthly precipitation had opposite patterns and different conditions of airflows. The causes and mechanisms were considered in connection with the state of the South Asian anticyclone at the 100 mb level. The principal conclusions are summarized as follows:

In spite of the different precipitation distributions in the respective years over South and Southeast Asia, the moisture distribution and the related process of occurrence of the SW monsoon do not differ greatly in each year. The occurrence of the SW monsoon can be explained by the occurrence and movement of the trough shown by the wind shear at the 850 mb level. It is suggested that the difference of rainfall over South and Southeast Asia may be caused by the difference of convection. The wet equatorial air mass followed by the equatorial westerlies changes into a southerly current in the vicinity of the Philippine Islands and travels further north over East Asia during the year with a small amount of precipitation. This significant airflow system seems to appear when the position of the east-west axis of the South Asian anticyclone is further south than normal and the tropical easterly jet stream is stronger at the 100 mb level over Southeast Asia. In particular, it should be noted that airflow patterns considerably different from the normal monsoon circulation can occur in some years.

References

Dao, S. -y., Chen L.-s. 1957: The structure of general circulation over continent of Asia in summer. *75th Ann. Vol. J. Met. Soc. Japan* 215–229.

Dao, S.-y. 1948: The mean surface air circulation over China. *Mem. Inst. Met. Academia Sinica* **15**(4).

Depperman, C. E. 1940: Upper air circulation (1–6 km) over the Philippines and adjacent regions. Philippines Weather Bureau, Manila. 64p.

Flohn, H. 1957; Large scale aspects of the summer monsoon in South and East Asia. *75th Ann. Vol. J. Met. Soc. Japan* 180–186.

Flohn, H. 1964: Investigations on the tropical easterly jet. *Bonner Met. Abh.* (4) 1–83.

Kao, Y.-h. 1948: General circulation of the lower atmosphere over the Far East. *Mem. Inst. Met. Academia Sinica* **16** (1).

Murakami, T. 1959: The general circulation and water-vapour balance over the Far East during the rainy season. *Geoph. Mag.* **29**(2) 131–171.

Nemoto, J., Kurashima, A., Yoshino, M. M., Numata, M. 1959: The Monsoons. Chijinshokan. 294p.

Ramanathan, K. R., Ramakrishnan, K. P. 1939: The Indian southwest monsoon and the structure of depressions associated with it. *Mem. India Met. Dep.* (26) 13–36.

Ramanathan, K. R. 1954: On upper tropospheric easterlies and the travel of monsoon and post-monsoon storms and depressions. *In* Proceeding of the UNESCO Symposium 243–244.

Ramdas, L. A. 1949: Rainfall of India, a brief review. *Indian J. Agricultural Science* **19** 1–19.

Ramdas, L. A. 1954: Prediction of the date of establishment of southwest monsoon along the west coast of India. *Indian J. Met. and Geoph.* **5** 305–314.

Riehl, H. 1954: Tropical Meteorology, London. 392p.

Sawyer, J. S. 1952: Memorandum on the intertropical front. *Met. Repts., British Meteorol. Office* **2** (10) 1–14.

Schmidt, F. H., Vecht, J. 1952: East monsoon fluctuations in Java and Madura during the period 1880–1940. *Verhandelingen*, Djawatan Meteorologie dan Geofisik of Indonesia 43.

Staff Members of the Section of Synoptic and Dynamic Meteorology, Institute of Geophysics and Meteorology, Academia Sinica, Peking. 1957: On the general circulation over Eastern Asia. *Tellus* **9** 432–446.

Subbaramyya, I., Ramanadham, R. 1966: The Asian summer monsoon circulation. *J. Met. Soc. Japan* **44** 167–172.

Thompson, B. W. 1951: An essay on the general circulation of the atmosphere over southeast Asia and West Japan. *Quart. J. Roy. Met. Soc.* **77** 569–597.

Wagner, A. 1931: Zur Aerologie des indischen Monsuns. *Gerl. Beitr. Geophy.* **30** 196–238.

Yin, M. T. 1949: A synoptic-aerological study of the onset of the summer monsoon over India and Burma. *J. Met.* **6** 394–400.

Yoshino, M. M. 1965, 1966: Four stages of the rainy season in early summer over East Asia. Part 1. *J. Met. Soc. Japan* **43** 231–245; Part 2. **44** 209–217.

Yoshino, M. M. 1969: Climatological studies on the polar frontal zones and the intertropical convergence zones over South, Southeast and East Asia. *Climatol. Notes, Hosei Univ.* (1) 1–71.

SOME ASPECTS OF THE INTERTROPICAL CONVERGENCE ZONES AND THE POLAR FRONTAL ZONES OVER MONSOON ASIA

Masatoshi M. YOSHINO

Abstract: The main frontal zone in Asia, the Pacific polar frontal zone, occurs south of Japan extending from SW–WSW to NE–ENE with frequencies of more than 20%. The latitudinal change of axis of the high frequencies ranges only about eight degrees from January to July. The other frontal zone, the eastern part of the Eurasian polar frontal zone, extends from Manchuria to the Sea of Japan with frequencies of about 15%. By analyzing the steadiness of the winds, the horizontal and the vertical structure of the intertropical convergence zones and the polar frontal zones over South, Southeast, and East Asia were examined. The intertropical convergence zones are apparently more complicated in summer than in winter. By contrast, the polar frontal zones are more complicated in winter. On the basis of cross-sections, it was suggested that active or marked intertropical convergence zones or polar frontal zones at the surface level are found when the zones of steadiness minimum at the upper levels are connected to them at the surface. Schematic illustrations of the intertropical convergence zones and the polar frontal zones were made for January and July by summarizing the results obtained in the preceding studies. Distribution maps of Bowen's ratio, sensible heat supply, wind velocity, and evaporation over the Northwest Pacific in January 1961, a normal winter, were made and the seasonal change along 145–150°E from 20° to 46°N was presented in comparison with the frequency of occurrence of the fronts. It was shown that in the westerly or northwesterly wind region over the Northwest Pacific in January, the zone with maximum amount of energy exchange of sensible heat appears first, and then, on the leeward side, the zone with the maximum amount of heat used for evaporation.

1. FREQUENCY OF OCCURRENCE OF POLAR FRONTAL ZONES OVER EAST ASIA

The frequency of occurrence of fronts is an important element in synoptic climatology. The distribution of fronts over the Hemispheres has been studied since 1930. In recent years, detailed studies of the frontal zones have been made from the standpoint of climatology for the whole Northern Hemisphere in winter and summer (Reed, 1960) and on annual changes in their distribution (Yoshimura, 1967). These studies revealed close relations between frontal zones and the sea surface temperature or the zonal wind speed cross-sections. Chu (1963) presented maps of frequencies of occurrence for the fronts over East Asia in January, April, July, and October, 1955–1959. His maps covered, however, China, Japan and their surrounding seas; that is, the region west of 145°E, excluding the main region of the northwest

Pacific. Climatological aspects of the frontal zones, the frequencies of occurrence from decade to decade in May, June, and July in a study of the Baiu, a rainy season occurring in early summer over East Asia, were examined at the surface and the 850 mb level in previous papers by the present writer (Yoshino, 1963, 1965, 1967a).

In the present study, the seasonal change in position of the polar frontal zones over the region between 120°E and 180°E in East Asia and the Northwest Pacific is first presented. Using daily weather maps of the surface over East Asia at 12.00 GMT, published by the Japan Meteorological Agency, fronts were counted in each area of two-degree latitude and longitude for each month from 1959 to 1963, regardless of the type of front, i.e. cold, warm, occluded or quasistationary fronts and frontogenesis or frontolysis. The five year total number of fronts in each square area was then divided by the number of maps used for the respective months and multiplied by 100 to give the distribution of the percentage frequency of occurrence. Latitude correction was employed for the values in each area.

a.   *Polar Frontal Zones at the Surface Level*

In January, the Pacific polar frontal zone runs clearly from south of Japan eastward along 32–33°N. In precise terms, the axis of the elongated region of high frequency is located at 30°N–130°E, 33°N–140°E, and 34°N–150°E.

In the figures presented by Reed (1960) and Yoshimura (1967), the Pacific polar frontal zone in January can be located far south of Japan, running ENE, or at 20°N–130°E, 22°N–140°E, and 25°N-150°E. The reason why the Pacific polar frontal zone appears about 10 degrees south in these figures seems to be because of the materials used, namely the Daily Series Synoptic Weather Maps published by the U.S. Weather Bureau.

In April, areas with a frequency greater than 20% are smaller. The principal frontal zone with a frequency greater than 15% begins from South China and runs ENE. On the other hand, the sub-frontal zone appears in Manchuria at 44–45°N. This zone can appropriately be described as the eastern part of the Eurasian polar frontal zone, which comes down from Siberia. In the figure presented by Chu (1963), the principal frontal zone in the Pacific and the intensified Eurasian polar frontal zone (b1, b2, and b3 in Fig. 2 of his paper) are also observed in April.

In July, a broad high frequency area extends eastward from North China to the Pacific, crossing the Japan Sea. This area covers more than 10 degrees latitude in width, from 32–33°N to 44–45°N. If one examines it in detail, three frontal zones can be seen keeping abreast with each other and extending SW–NE or WSW–ENE, the first in North China, the second in Korea and the Japan Sea, and the third on the southeast coast of Central Japan and the Northwest Pacific. The first frontal zone in China is also shown, in a slightly modified form, in the figure presented by Chu (1963). However, the second frontal zone mentioned above does not appear in his figure. The third appears at the same position in the figures presented by Reed (1960), Chu (1963), and Yoshimura (1967). This is a concentrated area of the Baiu front or the Pacific polar frontal zone. Synoptic climatological aspects of the frontal zone in the Baiu season have been studied in previous papers in relation to zonal wind characteristics at the 500 mb level (Yoshino, 1963, 1965). The most striking phenomenon in climatology in East Asia and the Northwest Pacific in summer must be this concentration, because of its high frequency and its length and size.

In October, a marked frontal zone develops over the southern coast of Japan, running ENE, while the frontal zone from Manchuria to the Japan Sea weakens. The autumn rainy season or *Shurin* in Japan occurs under these conditions. The figures given by Chu (1963) and Yoshimura (1967) also show clearly the development of a frontal zone at the same position.

To show the seasonal change in position, the concentrated areas of the frontal zones in January, April, July, and October are mapped in Fig. 1. Areas with a frequency of occurrence greater than 20% were determined to be the main frontal zones in the Pacific. Areas with a frequency greater than 15% on the other hand, were determined to be frontal zones over North China, Korea or the Japan Sea area because of their weaker concentration. These zones are called subfrontal. From Fig. 1, the following facts can be noted: 1) Although the main frontal zone shows a seasonal change in position, the range of south-north movement of its axis is not very great; between January and July it does not exceed eight degrees latitude. 2) In addition to the main frontal zone in the Pacific, there is a sub-frontal zone, as mentioned above. This zone is stronger in April and October than in January. 3) In July, the sub-frontal zone is also evident, but it is located southward. Consequently, the sub-frontal zone apparently extends to the main zone in a west-east direction across the Japanese islands.

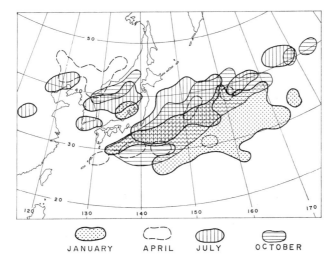

Fig. 1.   Seasonal change of position of the Pacific polar frontal zone at the surface level.

These phenomena seem to be related to the seasonal change in position of the jet stream over East Asia. The positional relations between the westerly jet streams and the Pacific polar frontal zones will be discussed in detail in a later section of this paper. However, generally speaking, in January, April, and October the sub-frontal zone, the eastern part of the Eurasian polar frontal zone, is located under the polar westerly jet stream which comes down from Siberia. By contrast, in July the sub-frontal zone is located under the subtropical westerly jet stream which extends to the northern side of the Himalaya-Tibetan Plateau in summer.

The subtropical and polar westerly jet streams flow together in the vicinity of Japan, attaining record wind velocities (Yoshino, 1967b). The position of the jet stream axis, however, changes only about ten degrees at 120°E, from 32°N in January to 42°N in July. In addition, it changes a little east of 140°E. In this connection, the main frontal zone in the Pacific changes its position only about eight degrees from January to July.

b. *Polar Frontal Zones at the 850 mb Level*

Generally speaking, the frontal zones at the 850 mb level appear in almost the same position west of 140°E in January: over the Japan Sea and the southern seas of Japan. On the other hand, the area with the highest frequency, i.e. 35%, is found in January at the 850 mb level but not at the surface level in the eastern seas of Northeast Japan.

The following can be said concerning the seasonal variation of the frontal zones at the 850 mb level in the region studied: 1) In January, the frontal zone develops strongly over southern Japan and extends to the Northwest Pacific. The zone width and its frequency of occurrence are also greatest in this month. 2) In April and October, a subfrontal zone appears to the east in Manchuria, in addition to the main frontal zone in the Pacific. The frontal zone in April resembles that in January. By contrast, in October it is closer in appearance to the July zone. 3) In July, the frontal zone moves north, and is apparently continuous from Manchuria to the Pacific.

2. Intertropical Convergence Zones and Polar Frontal Zones over South, Southeast, and East Asia as Revealed by Steadiness of the Winds

Over South, Southeast, and East Asia, the intertropical convergence zones are the most complex in the world. Many studies have been made of the location and vertical structures of the intertropical convergence zones by analyzing upper air data obtained since World War II. Recently, using satellite data, intertropical convergence zones were analyzed in detail (Sadler, 1963; Godshall, 1968; Hubert, 1965; Hubert *et al.*, 1969; Kornfield *et al.*, 1969; Sadler, 1969). These studies have shed new light upon the horizontal structure of these zones. In general, the studies agree on the locations of the intertropical convergence zones over India, while those over China run in a similar direction, SW to NE.

The zones are much more diverse over Southeast Asia than over the other two regions, South and East Asia. They can be divided into four general types: 1) those from the Indian Ocean crossing Sumatra to the Philippines; 2) those from India crossing Hainan Island; 3) those which run north-south near Taiwan or over southern Japan. This particular type has been studied by many researchers and been given various names, among which the most preferable appears to be the "Taiwan Convergence"; 4) those which run west-east over the Pacific. This is the western part of the main intertropical convergence zone continuing from the equatorial Pacific.

In winter the zones are not so diverse as in summer. The intertropical convergence zone runs west-east between 0–10°S at this time.

On the basis of the preceding results an attempt was made to assess the horizontal and vertical distribution of the intertropical convergence zones and the polar frontal zones from the climatological viewpoint; a low steadiness area is defined as a convergence or frontal zone. The steadiness values (1956–1960 averages) at every aerological station in these regions were obtained from Data 1 and 2. The definition and meaning of low values of steadiness were given in a previous paper (Yoshino, 1969a)

a. *Horizontal Distribution* (Fig. 2)

In January, the intertropical convergence zone is seen near the equator over Borneo and Celebes. The Southwest Asian polar frontal zone appears in West Pakistan at 30°N and then extends to the Bay of Bengal. Another branch is situated over Burma. The east part of the Eurasian polar frontal zone comes down along the southern side of Altai, between Mongolia

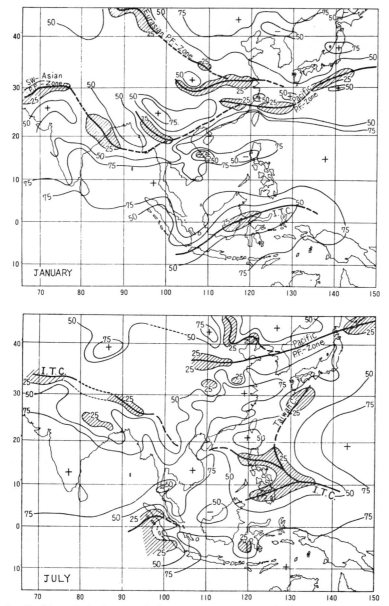

Fig. 2. Position of the intertropical convergence zones and the polar frontal zones in January and July.

and China, to Central China, where a branch of the Pacific polar frontal zone lies over the Yangtze Valley (30°N). The Pacific polar frontal zone starts from southern China, at about 25°N, and extends first eastward and then gradually northeast along the southern coast of Japan.

The westerly winds develop south of the intertropical convergence zone and the north-easterly winds to the north. These northeasterly winds are called the winter monsoon over

South and Southeast Asia, but trade winds over the Northwest Pacific. Between the monsoon and the trade winds, no marked boundary is found as far as the wind steadiness is concerned. However, the divergence field map for January published by Sandoval (1967) shows the convergence area over the Philippines. This implies that the boundary might be located over the Philippines. The northwesterly or westerly develops north of the Pacific ploar frontal zone, and brings the winter monsoon to northern China, Korea, and Japan. The low steadiness areas over India and Burma have been shown to be areas with a relatively sharp turn of air streams as illustrated by Wagner (1931). The convergence zone in January revealed by the resultant wind divergence fields calculated by Romanova and Romanov (1965) and Romanov (1965) coincides roughly with the position of the convergence over the tropical Indian Ocean given in Fig. 2.

In April, the intertropical convergence zone develops from the southern part of New Guinea and Celebes, via Malaya, to the Andaman Sea. The Southwest Asian polar frontal zone occupies northern India, crossing East Pakistan and extending to Vietnam. The eastern part of the Eurasian polar frontal zone appears north of 40°N, the same location as it has in January. The Pacific polar frontal zone expands, showing a double zonal structure, which was also revealed in the distribution pattern of the high value of rainfall in June (Yoshino, 1963).

The stream line shows the easterly to southeasterly at the southern side of the intertropical convergence zone and the easterly over the Northwest Pacific between the intertropical convergence and the Pacific polar frontal zone. The westerly develops both on the south side of the Southwest Asian polar frontal zone and the north side of the Eurasian polar frontal zone.

In July, the intertropical convergence zone in northern India, extends ESE to the Philippines. This is the northern intertropical convergence zone, if we classify the intertropical convergence zones into two: northern and southern ones. This northern intertropical convergence zone is cut down over north Vietnam, where the southwesterly, the summer monsoon over South and Southeast Asia, flows into East Asia bringing the rainy season in early summer. The southern intertropical convergence zone branches off from the main intertropical convergence zone in the Mindanao Islands, the Philippines, to Sumatra. However, it is also cut down in northern Borneo. A branch, the Taiwan Convergence, extending from east of the Philippines and Taiwan to western Japan, is also seen. It is very interesting that these three convergence zones were found separately, as mentioned above, but their existence has been checked climatologically, i.e. there are three convergence zones branching off in the Philippines, the SITC, the NITC and the Taiwan Convergence. It should be noted that the former two are obscured or perfectly cut down to the west of the Philippines. These locations are ascertained also by the divergence field presented by Sandoval (1967), the NITC over the equatorial Pacific connected mainly with the Taiwan Convergence and SITC over New Guinea and Celebes.

The stream line over South Asia is uniformly southwesterly. In East Pakistan it changes direction from southerly to southeasterly forming the monsoon trough in the Gangis Valley, northern India. The two low steadiness areas in eastern India have been clearly analyzed as boundary lines on the map presented by Wagner (1931), showing the SW monsoon at the 1 km level.

The easterly or eastsoutheasterly, the monsoon, in the area south of the equator, east of Java, Indonesia, flows into the Borneo Sea, then to the South China Sea, changing direction from southerly to southwesterly. After a confluence of the two southwesterlies, one from

South Asia and the other originating as the easterly from Indonesia and northern Australia, the southwesterly reaches western Japan via South China.

It has been pointed out from statistical investigation that the start of the rainy season in Japan in early summer is closely related not only to that of the SW monsoon in India, but also to that of the dry season in Java, caused by the E monsoon (Arakawa, 1959). This evidence can be explained by the fact that the cutting down of the southern intertropical convergence zone and the northern intertropical convergence zone makes possible the confluence of both SW air streams and the advance of the southwesterly to Japan. Arakawa's finding suggests that this circulation pattern takes place in parallel each year.

The Taiwan Convergence is caused by the conflict of two air streams, the southwesterly mentioned above and the southerly or southeasterly from the subtropical anticyclone over the Northwest Pacific. It had little frontal activity (Thompson, 1951), but it should not be neglected in the frontology of East Asia (Yoshino, 1963).

In October, the northern intertropical convergence zone runs at about 10°N over Southeast Asia. The southern intertropical convergence zone is not apparent, but is only traceable near Timor. The Southwest Asian polar frontal zone appears in northern India. The Eurasian polar frontal zone and the Pacific polar frontal zone come down to the same location as for January, but about 10 degrees north.

The southwesterly develop south of the northern intertropical convergence zone and the northeasterly north of it, just as in January. Over South Asia, the NE stream lines are not yet apparent, because the width between the northern intertropical convergence zone and the polar frontal zone is not very great. Over the Indian Ocean, according to observations at the 4 km level in October 1962, it was shown that the NITC runs at about 5°N from south of Ceylon to north of Sumatra and the SITC from the point 10°S, 75°E, to the point 2°S, 95°E (Lukyanov, 1965), which differs from the results obtained previously.

b. *Vertical Structure* (Figs. 3(a) and 3(b))

The vertical distributions of steadiness are given for January and July at the cross-sections along 80, 100, 120 and 140°E respectively. The lines showing the steadiness-minimum and the positions and velocity of the jet streams are presented as well as the steadiness distribution.

The line of steadiness-minimum must be described, because its seasonal shifting is very clear and is closely related to the characteristics of the intertropical convergence zone or the polar frontal zone. In January, along 80°E the westerly jet develops markedly (90 knots) at 25°N, below it, the steadiness-minimum line, starting from the 30,000 ft level, reaches the surface, where the South Asian polar frontal zone develops at the surface level. Along 100°E, the steadiness-minimum line comes down to 20°N at the surface, where the South Asian polar frontal zone develops just below the westerly jet. The northern intertropical convergence zone is located near the equator in January and April, but its range is small. In contrast to the obscured Pacific polar frontal zone on the southern side of the Tibetan Plateau in January and April, the Eurasian polar frontal zone on the northern side is apparent. Along 120°E, the westerly jet, the strongest in the world, is located at the 150 mb level at 32°N and has a velocity of 130 knots in January. Below it, the steadiness-minimum line reaches the surface, where the polar frontal zone develops. Along 140°E, the distribution is something like a schematic figure in January: The westerly south of the equator, the easterly between the intertropical convergence zone and the steadiness-minimum line, the westerly at a high latitude north of the line, and the low steadiness are between

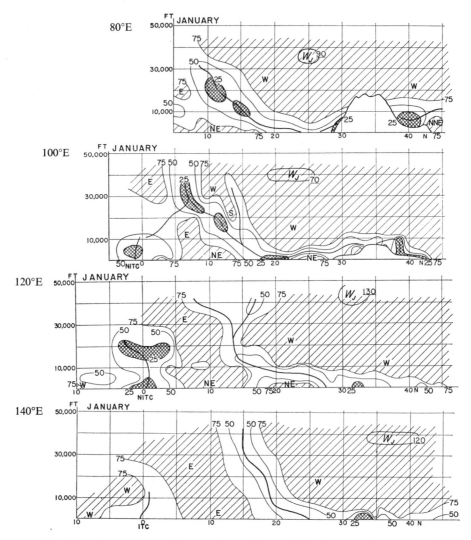

Fig. 3(a). Cross-sections of the steadiness of the winds in January. W: westerly wind, E: easterly wind, etc. W_j: axis of westerly jet stream, E_j: axis of easterly jet stream. Values underlined: mean wind velocity in knots. Thick line: steadiness-minimum line.

30°N, which corresponds to the Pacific polar frontal zone below the westerly jet stream.

In July the steadiness-minimum line between the easterly and westerly moves north of the Himalayas. At a higher level, the westerly occupies the northern side of this line in contrast to the easterly on the southern side. The height of the northern intertropical convergence zone seems to be influenced by the widely developed easterly at the upper tropospehre south of 30°N. It is said by mountain climbers that when the summer monsoon starts in India, unstable weather appears in the Himalayas. This saying was first explained by this figure from a synoptic climatological viewpoint: that is, the start of the SW monsoon over India, even though its northern limit and height is far from the Himalayas, means the northward shifting of the polar frontal zone associated with the steadiness-minimum line and

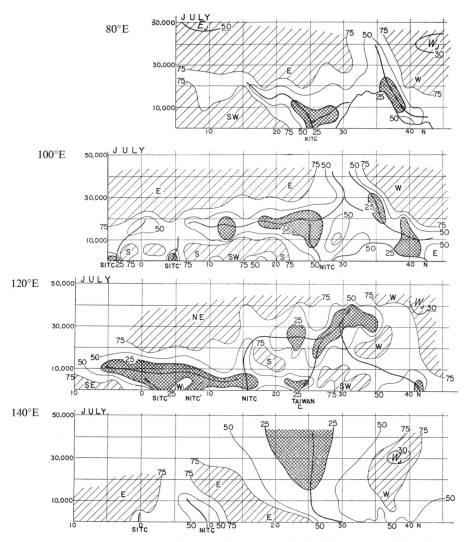

Fig. 3(b). Cross-sections of the steadiness of the wind in July. Same as (a) but in July.

consequently the polar frontal zone comes just over the Himalayas, bringing bad weather. Along 100°E, the structure is quite striking in July. The SITC and its branch, the SITC′, are located near the equator and between these two and the NITC, the southerly or southwesterly region is occupied up to the 700 mb level. Above this region a thick layer with low steadiness is seen between the equator and 25°N above the northern intertropical convergence zone. This is considered to be a result of greater variation, occuring from day to day and year by year, in the height of the upper limit of the southerly and southwesterly, or summer monsoon over Southeast Asia, or the lower limit of the easterly originating from the South Asian high at the upper troposphere. These characteristics are also seen on the sections along 80°E and 120°E, but along 100°E they are most apparent. That is to say, the southwesterly at the lower troposphere is most variable along 100°E in July.

The distribution pattern along 120°E in July is the most complicated. The height of the SITC, the NITC′, the NITC and the Taiwan Convergence decreases from 10,000 ft at the equator to 5,000 ft at 15°N. Between the NITC and the Taiwan Convergence, the southerly appears at higher levels and between the Taiwan Convergence and the Pacific polar frontal zone at 42°N, the southwesterly prevails below 10,000 ft. Between the northeasterly and the westerly at the upper troposphere, there is a large area of low steadiness. The most striking feature is the low steadiness area between 20 and 28°N in July and between 10 and 25°N in October at the upper troposphere above the 500 mb level. It should be pointed out that the easterly develops below this area. This fact suggests the weakness, variability and structure of the anti-trade winds in connection with meridional circulation (Yoshino, 1967b).

3. Schema of the Intertropical Convergence Zones and the Polar Frontal Zones over South, Southeast, and East Asia

In this section schematic illustrations of the intertropical convergence zones and the polar frontal zones over South, Southeast, and East Asia at the 850 mb level are given for January and July. The schema were drawn from data presented in previous studies and in the preceding parts of this study.

For the polar frontal zones surrounding Japan and the Northwest Pacific, reference was made primarily to Figs. 1 and 2 of the present paper and for China, to Figs. 1 and 3 of the paper by Chu (1963) and Fig. 2 of this paper. For the polar frontal zones over South Asia, Fig. 2 of this paper was used principally.

The locations of the intertropical convergence zones were determined by referring first to Fig. 2 and secondly to the following data obtained in recent years: 1) Maps of the monthly frequency of the intertropical convergence zones by Yoshimura (1970) obtained from the Synoptic Weather Maps for the tropical region during the IGY, 1958, published by the Deutscher Wetterdienst, Seawetteramt, Hamburg; 2) the mean surface position of the intertropical convergence zone determined by surface wind constancy and total cloud amount obtained from TIROS nephanalyses for the months of January 1963, 1964, and 1965 (Godshall, 1968); 3) multiple exposure satellite pictures which show the monthly averages from ESSA III and ESSA V for 1967 (Kornfield and Hasler, 1969); 4) the seasonal distribution of brightness from ESSA III and ESSA V for the period from March 1967 to February 1968 (Hubert et al., 1969); 5) maps showing the divergence field and the streamlines and convergence zone of the mean surface winds for each month (Sandoval, 1967); and 6) the monthly cloudiness distributions of the two year average for 1965 and 1966 from satellite pictures (Sadler, 1969).

Statistically speaking, the data from these maps are of course heterogeneous with respect to climatological materials: they were obtained for different periods, based upon various methods of determining intertropical convergence zones, and given for the surface level, the 850 mb level or certain cloud levels. However, they are more reliable than previous results, which were established on scarce observation points in the equatorial region. In addition, the streamlines were drawn on the schema with reference to maps of the monthly mean resultant winds at the 850 mb level over the Pacific (Wiederanders, 1961) and at the 2,000 ft level over Southeast Asia (Data 1 and 3).

a. *Schema in January*

Figure 4(a) is the schema in January. Broadly speaking, the intertropical convergence zone is situated between the equator and 10°S. Some parts in this zone develop intensively, but they no longer form a belt-like region. In the region of central Celebes and southeastern Borneo and the Pacific east of 150°E, the northern intertropical convergence zone appears along the equator. The northern intertropical convergence zone in the Pacific was detected in the satellite pictures, although it is not very wide or intensive. On the southern side of the intertropical convergence zone, southwesterly airstreams develop. On the other hand, the northeasterly winds known as the winter monsoons over South and Southeast Asia and the trade winds in the Pacific.

The polar frontal zones are quite complex, with many branches markedly developed. Generally speaking, the location of the Southwest Asian, South Asian, and Pacific polar frontal zones coincides with that of the subtropical westerly jet stream, which reaches record wind speeds over East Asia in January. On the northern side of the polar frontal zones develop the northwesterly winds known as the winter monsoon in North China, Korea, and Japan. An interesting question, from the synoptic climatological viewpoint, is the position of the Pacific polar frontal zone in the westerly or northwesterly air stream region in the Pacific. Further study is needed to clarify this point in detail.

b. *Schema in July*

The schema in July is given in Fig. 4(b). Generally speaking, the distribution of the polar frontal zones is relatively simple in contrast to the complex situation of the intertropical convergence zones. The southern intertropical convergence zone comes up from the southeast in the South Pacific to the Solomon Islands and New Guinea. The extended parts of the southern intertropical convergence zone can be traced in Celebes and as far as South Sumatra.

On the southern side of the southern intertropical convergence zones, eastsoutheasterly winds are predominant. These winds take a sharp turn to the right after crossing the equatorial region and blow into South and Southeast Asia as the SW summer monsoon. No intertropical convergence zone is found in the Indian Ocean. This enables the SW air streams to transport a large quantity of water vapor over South and Southeast Asia in summer and to produce heavy rainfall over East Asia during the rainy season in early summer in accordance with the inflow of the airstreams there.

Along the southern border of the Tibetan Plateau and the mountainous region of Yunnan in China, the northern intertropical convergence zone is located. In the equatorial Pacific in the Northern Hemisphere, on the other hand, the clearly developed northern intertropical convergence zone is located between 5°N and 10°N east of 145°E, and then gradually goes north and branches off into three zones east of the Philippine Islands: the Taiwan convergence, the northern intertropical convergence zone (NITC) over Luzon and the NITC' over Mindanao to northern Borneo.

An anticyclonic circulation develops north of the northern intertropical convergence zone and east of the Taiwan convergence in the Pacific. The frontal zone along the Yangtze River in China is recognized from its nature as a tropical frontal zone, as pointed out by Kurashima (1959), but from its location it is also recognized as the western part of the Pacific polar frontal zone. In any case, this frontal zone over China shows many different characteristics from the Pacific polar frontal zone near Japan as studied in the previous paper (Yoshino, 1963).

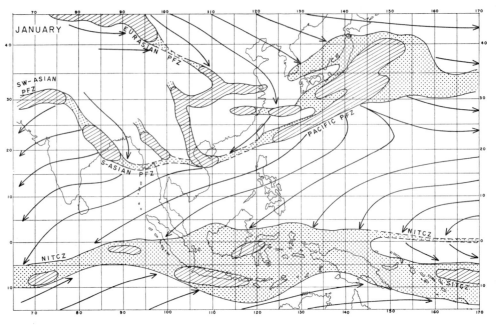

(a) January.

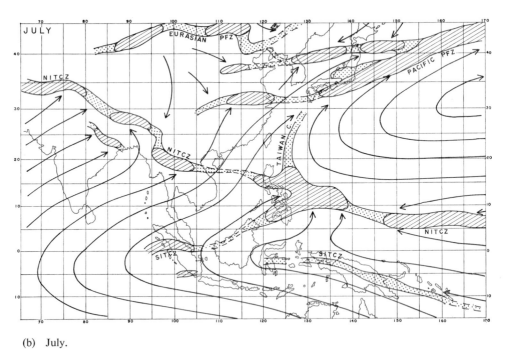

(b) July.

Fig. 4. Schemata of the intertropical convergence zones and the polar frontal zones over South, Southeast, and East Asia in January and July.

4. HEAT EXCHANGE AND THE POLAR FRONTAL ZONE OVER THE NORTHWEST PACIFIC

In recent years, the heat balance of the Japan Sea has been studied by many scientists (Miyazaki, 1949; Aldoshina, 1957; Manabe, 1957, 1958; Fujita and Honda, 1965; Nino-miya, 1968). The methods and the results of these studies are applicable to the frontogenesis problems over the Northwest Pacific.

As given in Fig. 4(a), the Pacific polar frontal zone is located in the westerly or north-westerly region south of Japan in winter. This is in contradiction with the general definition or recognition of fronts in the meteorological sense. This section will, therefore, clarify this fact and explain its causes from the viewpoint of heat exchange.

For this study marine climatological tables of the North Pacific (Data 4) were used. The tables, based on ship observations, cover the area 0–60°N and 100°E–170°W, divided into squares of 2° latitude and 5° longitude. For each square, the monthly means of air tempera-ture T_A, sea surface temperature T_W, dew-point temperature, wind velocity, air pressure etc. are presented.

First, Bowen's ratio, R, is calculated by

$$R = 0.64 \frac{P}{1,000} \frac{T_W - T_A}{e_W - e_A},$$ (1)

where P: air pressure (mb), T_W: sea surface temperature (°C), T_A: air temperature (°C), e_W: saturated water vapor pressure (mb) at temperature of T_W, and e_A: water vapor pressure (mb) of air temperature. The sensible heat supply (ly/day), Q_h, is obtained by Jacobs' formula (Jacobs, 1951),

$$Q_h = 5.5(T_W - T_A) v,$$ (2)

where v is wind velocity in m/s. The evaporation (mm/day), E, is given by

$$E = 0.143 (e_W - e_A) v.$$ (3)

Distribution maps of Bowen's ratio, wind velocity, sensible heat supply and evaporation were made for January, 1961, a normal winter, and for January, 1963, an example of an anomalous winter. Furthermore, to show the seasonal variation, these terms were obtained for January, April, July, and October for the five year average of 1959–1963 along the meridi-onal zone of 145–150°E.

a. *Bowen's Ratio*

The distribution of Bowen's ratio in January, 1961, shows a striking regionality, as given in Fig. 5. In the zonal region of the Northwest Pacific between 40°N and 50°N west of 170°W as well as in the Japan Sea and the Yellow Sea, values greater than 0.6 occur. The greatest value, 1.9, is found in the region east of Hokkaido in northern Japan. According to the distribution map of Bowen's ratio as calculated for normal (Fujita and Honda, 1965), the greatest value, 1.8, is found at the coast near Vladivostok in the Japan Sea. Manabe has calculated also the distribution of Bowen's ratio in the Japan Sea for the outburst period of the winter monsoon at the end of December 1954 (Manabe, 1957) and for the whole winter season from December 1954 to February 1955 (Manabe, 1958). In these maps values greater than 1.5 appeared in the coastal region of Vladivostok. The isoline 1.0 runs from the Tongjoson Gulf, North Korea, to the southwestern part of Hokkaido, northern Japan, roughly along 40°N. As a normal condition, therefore, it is suggested, in summarizing the results mentioned above, that the isoline 1.0 continues from 39°N on the east coast of North Korea to 42°N on the west coast of Hokkaido and then from 43°N on the east coast of Hokkaido to further east. Much greater values are seen north of this line.

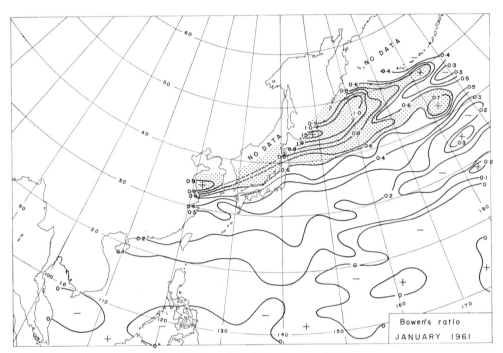

Fig. 5.   Bowen's ratio in January, 1961.

b.   *Wind Velocity*

The distribution of wind velocity is given for the normal winter of January, 1961, in Fig. 6. In the northernmost part of the Northwest Pacific, the wind attains a velocity of over 15 m/s. The wind maximum zone starts at the southern coast of central Japan and runs eastwards with the axis reaching wind velocities of 15 m/s at certain places. In parallel with this, the axes of evaporation maximum and sensible heat supply are seen in Figs. 7 and 8. The axis of the maximum occurrence of fronts, the frontal zone, runs parallel with this axis, but situates about 2° latitudinally south. According to Fujita and Honda (1965), the wind velocity is over 18 m/s in a small area in the northern part of the Japan Sea in January, but 8–12 m/s over most of the Japan Sea except for the coastal regions.

c.   *Sensible Heat Supply*

Figure 7 presents the distribution of sensible heat supply in January, 1961. Values greater than 400 ly/day appear in a zone from the Yellow Sea via the Japan Sea to the Northwest Pacific. The distribution of total heat exchange in January given by Wyrtki (1966) shows the same pattern. He calculated the total heat, Q_t, received or lost by the ocean at the sea surface by the formula,

$$Q_t = Q_i - Q_b - Q_e - Q_h, \tag{4}$$

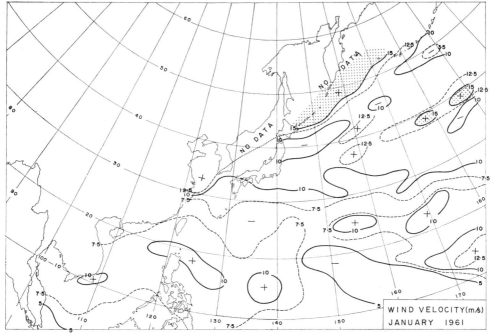

Fig. 6. Wind velocity in January, 1961.

where Q_i is the incoming radiation from sun and sky, Q_b the effective back radiation from the sea surface, Q_e the heat lost in evaporation and Q_h the exchange of sensible heat between ocean and atmosphere. In the zonal area from the southern coast of Japan eastward, there appear values of more than 800 ly/day of the total heat loss of the ocean. A zone with values of 400 ly/day extends as far as 175°E. Comparing Wyrtki's results with Fig. 7, the most striking feature apparent over the Northwest Pacific might be this zonal concentration. The annual as well as winter values of the total heat loss from the sea surface in this zone are, however, smaller than for the similar zonal area developed within the Gulf Stream over the North Atlantic, as has been pointed out by Jacobs (1951). In a study on the daily heat exchange in the North Pacific (Laevastu, 1965), it was indicated that the air gained the highest sensible heat, more than 800 ly/day, in the northern part of Kamchatka on February 14, 1957.

The distribution of the sensible heat supply in the anomalous winter in January, 1963, revealed that an area with values greater than 400 ly/day develops around Japan. The highest values, more than 800 ly/day, occurred in the Japan Sea. The sensible heat supply during the period of intense outburst of the winter monsoon reached as much as 1,030 ly/day (Manabe, 1957) and 1,100 ly/day (Aldoshina, 1957) in the Japan Sea. During the ten-day period of January 16–25, 1963, when the winter monsoon was the strongest in recent years, the amount actually reached 1,000 ly/day in the area south of Vladivostok (Fujita and Honda, 1965). As a monthly mean, the sensible heat supply from the whole Japan Sea area was 555 ly/day for January and February in 1955 (Manabe, 1958) and 740 ly/day in January, 1963, in contrast to 340 ly/day in 1964 or 410 ly/day in 1965 (Ninomiya, 1968). Therefore, it can be seen that the sensible heat supply in an anomalous winter was a little less than twice that in a mild winter.

An interesting point to be noted is the absence of an obvious zone in the Pacific in January, 1963. This implies that the sensible heat supply was not higher than normal in the Pacific east of Japan during this month, in contrast to the high values in the Japan Sea or south of Japan.

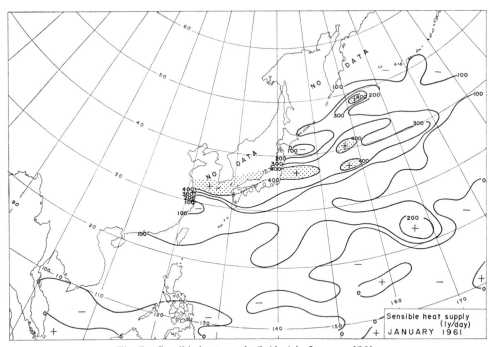

Fig. 7.   Sensible heat supply (ly/day) in January, 1961.

d.   *Evaporation*

The distribution of evaporation in January, 1961, as given in Fig. 8, is relatively complex. A zone with a high evaporation of more than 12.5 mm/day develops from the southern coast of Japan eastward, located in the same position as the zone of sensible heat supply shown in Fig. 7. In addition, one can observe a second zone with its axis along 28–29°N east of 166°E and a third zone with its axis fluctuating between 21 and 24°N east of 150° E. The third zone coincides approximately with that of wind velocity given in Fig. 6. In general, the distribution pattern of evaporation resembles the pattern of wind velocity south of 25°N in the Northwest Pacific.

The distribution of evaporation in January, 1963, shows a quite different pattern than that in January, 1961. It has, of course, a strong resemblance to the wind velocity pattern. Interestingly, the area with a great amount of evaporation shifted south-eastwards compared to the area with high values of sensible heat supply.

Using Jacobs' formula, the evaporation from the whole Japan Sea for January was calculated as 10.6 mm/day in 1963, 6.8 mm/day in 1964 and 8.0 mm/day in 1965 (Ninomiya, 1968). Recently, an attempt was made to devise an evaporimeter at the sea surface (Harami, 1969). The experimenter found, through an experimental formula based upon four-year observations at 75 points in the Japan Sea, that a maximum value of 14 mm/day was seen in

January, 1968, in the area 150–200 km west of Akita City, Tohoku District, Japan, where the wind was strong and the sea surface warm at the same time.

In the severe winter of January, 1963, the maximum was greater: i.e. 16 mm/day at a point 200 km west of Akita City and at the northern part of the Tsushima Strait, according to calculations by Fujita and Honda (1965). But, the maximum was found at the southern coast of Japan with a value of more than 20 mm/day.

The average monthly amount of heat lost in evaporation in January is greatest in the Pacific: the narrow, short zonal area east of the Bôsô Peninsula, Kantô District, Japan, has a value of over 600 ly/day, or approximately 10.4 mm/day of evaporation.

It can be concluded, therefore, that the evaporation from the Japan Sea in January of a normal winter is 6–7 mm/day, while the zonal area extending from the Kantô District east has 10 mm/day or more. But in the unusually severe winter recorded, the maximum is almost twice as large although the areas with a maximum are shifted or limited.

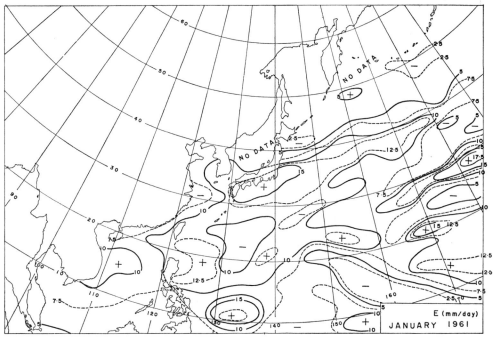

Fig. 8. Evaporation (mm/day) in January, 1961.

e. *Seasonal Variation*

The seasonal variation of Bowen's ratio, wind velocity, sensible heat supply and evaporation as well as the difference between the water and air temperature along 145–150°E for each 2 degrees latitude from 20°N to 46°N is given in Table I. For comparison, the frequencies of occurrence of the fronts are also compiled in the table.

In contrast to the marked meridional shifting of the maximum zone of the frequencies of occurrence of the fronts, the other terms presented in Table I do not show any obvious seasonal change except for wind velocity. Bowen's ratio is at a maximum 1.03 at 42–44°N in January. Around this zone, values greater than 0.6 occur in accordance with zones of higher wind velocity north of 40°N in January. In addition to this wind maximum in the

Table I(a). Bowen's ratio, wind velocity and sensible heat supply along 145–150°E for the five-year average, 1959–1963.

	Bowen's ratio				Wind velocity (m/s)				Sensible heat supply (ly/day)			
	Jan.	Apr.	July	Oct.	Jan.	Apr.	July	Oct.	Jan.	Apr.	July	Oct.
44–46°N	0.56	0.20	0.69	0.20	22.3	14.6	11.1	16.2	161	26	−52	64
42–44°	1.03	0.64	−0.39	0.14	20.1	10.0	10.9	16.6	152	−22	−17	18
40–42°	0.62	−0.10	−1.62	0.53	25.6	14.8	10.2	15.1	169	13	−45	50
38–40°	0.58	0.06	−0.38	0.14	19.1	16.0	12.3	15.7	280	32	−25	59
36–38°	0.51	0.16	−0.40	0.17	16.5	17.0	11.1	17.4	318	61	−34	94
34–36°	0.37	0.11	−0.14	0.12	20.9	17.5	12.0	18.2	283	44	−20	72
32–34°	0.28	−0.05	−0.13	0.09	22.7	17.7	11.5	18.4	198	27	−28	50
30–32°	0.20	−0.11	−0.03	−0.05	21.8	17.6	11.5	11.7	157	19	−1	−14
28–30°	0.11	−0.06	−0.06	(0.07)	21.4	12.3	7.9	13.4	110	12	−15	(26)
26–28°	0.11	−0.16	−0.17	−0.17	17.6	10.5	6.8	11.1	70	6	−8	−31
24–26°	0.12	−0.24	0.05	−0.16	10.7	12.2	7.4	16.7	66	−62	13	−52
22–24°	0.09	−0.35	0.02	−0.08	12.1	13.3	9.1	14.4	52	−27	7	−29
20–22°	0.07	−0.04	0.02	−0.05	11.4	13.2	10.3	15.9	33	−4	5	−25

Table I(b). Evaporation, difference between water and air temperature, and occurrence frequency of fronts along 145–150°E for the five year average, 1959–1963.

	Evaporation (mm/day)				$T_W - T_A$ (°C)				Occur. freq. fronts (%)			
	Jan.	Apr.	July	Oct.	Jan.	Apr.	July	Oct.	Jan.	Apr.	July	Oct.
44–46°N	1.8	−0.4	−1.1	4.2	3.5	−0.5	−2.0	0.1	10	9	13	12
42–44°	2.5	0.7	0.0	3.0	2.0	−1.4	−1.3	0.2	11	12	18	11
40–42°	5.0	2.1	0.2	5.5	3.2	−0.2	−1.5	1.0	13	14	21	12
38–40°	8.1	3.8	2.1	6.9	5.1	0.7	−0.7	1.4	16	15	21	19
36–38°	10.7	5.7	1.7	8.5	5.6	1.0	−1.1	1.5	18	18	23	23
34–36°	13.0	6.3	3.1	10.1	4.7	0.9	−0.5	1.4	26	19	23	27
32–34°	13.0	4.8	4.3	9.9	3.4	0.3	−1.0	1.0	27	18	20	30
30–32°	13.0	4.6	5.4	4.1	2.5	−0.3	−0.1	−0.4	23	19	15	23
28–30°	11.0	3.8	3.1	7.3	1.1	0.3	−0.5	−0.1	22	17	7	16
26–28°	12.8	2.7	2.5	3.3	1.5	0.1	0.1	−0.8	19	15	4	13
24–26°	8.8	1.7	4.4	5.9	2.0	−1.4	0.6	−0.5	16	13	1	11
22–24°	9.1	4.1	4.8	7.9	1.4	−0.6	0.1	−0.6	15	11	0	7
20–22°	7.7	5.6	6.1	8.2	0.8	−0.1	0.2	−0.6	12	7	0	4

north, another wind maximum zone exists between 28–36°N in January, 30–38°N in April, 30–40°N in July, and 32–38°N in October.

The sensible heat supply shows no clear seasonal variation in position. A seasonal change in absolute values can be observed, however: the value of 318 ly/day at 36–38°N in January is five times greater than that at the same latitude in April and three times that in October. This corresponds naturally with the seasonal variation in the difference between water and air temperature.

The amount of evaporation shows a similar tendency for seasonal variation as the sensible heat supply, but the degree of variation between winter and summer is less. Moreover, the positional change of the evaporation maximum is much closer to the wind velocity. The maximum evaporation reached 13 mm/day, or approximately 750 ly/day.

The concentrated zone of the occurrence frequency of the fronts is 28–36°N in January,

30–38°N in April, 32–42°N in July, and 30–38°N in October. In January, the most significant month, the zone with maximum occurrence frequency of the fronts coincides perfectly with that of wind velocity. At about four degrees north of this zone, the maximum sensible heat supply is located. The zone of maximum evaporation is also located at the same latitude or two degrees north of this zone. Therefore it can be said that, considering the prevailing wind direction given in Fig. 4(a), the zone with the maximum amount of heat exchange of sensible heat appears first and then the zone with the maximum amount of heat lost in evaporation in the northwesterly or westerly region. Further to leeward, the zone of maximum frequency of occurrence of the fronts is seen.

Of course, this is not the case in July. In April and October, the same pattern is found, even though the values and ranges are smaller.

f. *Relation between the Position of Intertropical Convergence Zones and Polar Frontal Zones and the Sea Surface Temperature*

In the preceding section, it was shown that the position of the frontal zone is closely related to the zones of maximum sensible heat supply and maximum amount of heat lost in evaporation. In simpler terms, it could also be said that the position of the frontal zone is related to the zone of maximum difference between sea surface and air temperature. This is due to the fact that the zones of maximum sensible heat supply and maximum amount of heat lost in evaporation are located paralleled to the zone of maximum difference between the sea surface and the air temperature, as was shown in Table I. At this point, the relation between the positions of the intertropical convergence zones or the polar frontal zones and the difference between the sea surface and the air temperature over East and Southeast Asia will be reexamined.

The average values for 1956–1960 of the sea surface temperature and the difference between the sea surface and the air temperature were calculated to study their relationship to the positions of the intertropical convergence zones and the polar frontal zones.

The maps showing the distribution of the sea surface temperature and the areas where the sea surface is warmer than the air in January and in July were omitted due to lack of space. It can be pointed out, however, that as far as the area over the Northwest Pacific is concerned, perfect coincidence is observed between the position of the Pacific polar frontal zone and the area where the sea surface is 4–7° warmer than the air. It has been pointed out by Reed (1960) that the position of the polar frontal zone in the Pacific is related to the sea surface temperature distribution. However, the distribution of the temperature difference between the sea surface and the air may better explain the position of the polar frontal zone through the process of air-sea interaction.

The figures in July again accord well with each other. The position of the intertropical convergence zone coincides with the area east of the Philippines where the sea surface is 1–3°C warmer than the air. The Taiwan convergence and the Pacific polar frontal zone in eastern Hokkaido, Japan, are clearly seen also in the warmer sea surface region. From the climatological viewpoint, therefore, the positions of the intertropical convergence zone and the polar frontal zone accord with the areas where the sea surface is more than 4°C warmer than the air in January and 1–3°C in July.

In a study of the circulation over the eastern equatorial Pacific, it was indicated that the sea surface temperatures are particularly important in heat convection when they are greater than 28–29°C (Krueger and Gray, 1969). In the present paper the vertical structure is not made clear, but the conditions mentioned above suggest the important role that sea surface temperatures play in the formation of the intertropical convergence zone also over the western equatorial Pacific.

5. SUMMARY

Climatological studies of the polar frontal and intertropical convergence zones over South, Southeast, and East Asia have revealed the following facts:

1) The main frontal zone, the Pacific polar frontal zone, develops along the southern coast of Japan from SW–WSW to NE–ENE with occurrence frequencies greater than 20%.

2) The latitudinal change of the main frontal zone is only about eight degrees from January to July. In January, the frontal zone develops markedly and has the greatest width and frequency.

3) The subfrontal zone, the eastern part of the Eurasian polar frontal zone, occurs from Manchuria to the Sea of Japan with frequencies of about 15%. This zone was evident in April and October.

4) The main frontal zone extends to the north, apparently continuing from Manchuria to the Northwest Pacific.

5) The structure of the intertropical convergence zones is more complex in summer than in winter over Southeast Asia.

6) In July, the intertropical convergence zone branches over the Philippines into: i) the Taiwan Convergence extending north, ii) the northern intertropical convergence zone extending to Hainan Island, and iii) the southern intertropical convergence zone extending to the northern part of Borneo.

7) The structure of the polar frontal zones was more complex in winter. They occupied the Southwest Asian polar frontal zone over northern India with some branches, the eastern part of the Eurasian polar frontal zone over China and the Pacific polar frontal zone from the East China Sea to the southern coast of Japan.

8) According to vertical cross-sections along 80, 100, 120, and 140°E, it was suggested that the active or apparent intertropical convergence zones and polar frontal zones at the surface level were found where the steadiness minimum line reaches to the surface.

9) The height of the intertropical convergence zones was lower in summer, probably due to the strong easterly at the upper level. In many cases, the northern intertropical convergence zone was broad at the upper part.

10) Along the intertropical convergence zones, the frequency of occurrence of rainfall is relatively high (20–30%), and occasionally more than 250 mm with a maximum greater than 400 mm in July.

11) Along the Pacific polar frontal zone, the Baiu front, the monthly rainfall reaches more than 500 mm in Central China and Western Japan in July.

12) Along the subfrontal zone, a belt with rainfall of more than 500 mm prevails from the central part of the Korean Peninsula to Western Japan in July.

13) Schematic illustrations of the intertropical convergence zones and polar frontal zones were presented with the streamlines for January and July. These can be used to interpret the distribution of precipitation and other climatic elements over the region.

14) The distributions of Bowen's ratio, sensible heat supply, wind velocity and evaporation over the Northwest Pacific were studied in January, 1961, as an example of a normal winter. It was indicated that in the westerly or northwesterly wind region over the Northwest Pacific, the zone with maximum amount of energy exchange of sensible heat appears

first and then, on the leeward, the zone with the maximum amount of heat lost in evaporation.

15) The positions of the intertropical convergence zone and the polar frontal zone accord with the areas where the sea surface is more than 4°C warmer than the air in January and 1–3°C in July.

16) The highest climatological amounts of heat exchang terms in January are a sensible heat supply of 800 ly/day in the Japan Sea and, for heat lost in evaporation, 750 ly/day (13 mm/day of evaporation) between 30 and 36°N along 145–150°E in the Northwest Pacific.

References*

Aldoshina, E. I. 1957: Teplovoĭ balans poverkhnosti Iaponskogo moria. *Trudy Gos. Okeanogr. Inst.* (Moscow) **35** 119–157.

Arakawa, H. 1959: Relation among beginning of dry season in Java, establishment of SW monsoon in India and onset of Bai-u in the Far East. *Proc. IGU Reg. Conf. Japan 1957* 589–592.

Chu, R.-ch. 1963: The climatic frontal zones over East Asia. *Acta Met. Sinica* **33** 527–536.

Fujita, T., Honda, T. 1965: Observational estimation of evaporation and sensible heat transfer from the Japan Sea in winter. *Tenki* **12**(6) 204–213 (in Japanese).

Godshall, F. A. 1968: Intertropical convergence zone and mean cloud amount in the tropical Pacific Ocean. *Mon. Wea. Rev.* **96**(3) 172–175.

Harami, K. 1969: On the evaporation from the Japan Sea. *Tenki* **16**(10) 469–475 (in Japanese).

Hubert, L. F. 1965: ITC and high level analysis aids. Proc. Inter-Regional Sem. on Interpretation and Use of Meteorological Satellite Data. *Japan Met. Agency Tech. Rep.* (47) 47–56.

Hubert, L. F., Krueger, A. F., Winston, J. S. 1969: The double intertropical convergence zone—fact or fiction? *Jour. Atmos. Sci.* **26**(4) 771–773.

Jacobs, W. C. 1951: Large-scale aspects of energy transformation over the oceans. *Comp. Met.* 1057–1070.

Kornfield, J., Hasler, A. F. 1969: A photographic summary of the earth's cloud cover for the year 1967. *Jour. Appl. Met.* **8**(4) 687–700.

Krueger, A. F., Gray, T. I. 1969: Long-term variations in equatorial circulation and rainfall. *Mon. Wea. Rev.* **97**(10) 700–711.

Kurashima, A. 1959: General circulation and monsoon. *In* Kisetsufû by J. Nemoto et al. 201–283 (in Japanese).

Laevastu, T. 1965: Daily heat exchange in the North Pacific, its relations to weather and its oceanographic consequences. Commentationes Physico-Mathematicae, Ed. Societas Scientiarum Fennica **31**(2) 1–53.

Lukyanov, V. V. 1965: Some features of the summer monsoon over the northeastern Indian Ocean. *Trudy Gos. Okeanogr. Inst.*, Acad. Nauk SSSR (78) 81–91.

Lukyanov, V. V., Romanov, Yu. A. 1965: On the intertropical convergence zone over the Indian Ocean. *Trudy Gos. Okeanogr. Inst.*, Acad. Nauk SSSR (78) 92–110.

Manabe, S. 1957: On the modification of air-mass over the Japan Sea when the outburst of cold air predominates. *Jour. Met. Soc. Japan II*, **35**(6) 311–326.

Manabe, S. 1958: On the estimation of energy exchange between the Japan Sea and the atmosphere during winter based upon the energy budget of both the atmosphere and the sea. *Jour. Met. Soc. Japan II*, **36**(4) 123–134.

Miyazaki, M. 1949: The incoming and outgoing heat at the sea surface along the Tsushima warm current. *Oceanogr. Mag.* **1** 103–113.

Ninomiya, K. 1968: Heat and water budget over the Japan Sea and the Japan Islands in winter season. *Jour. Met. Soc. Japan II*, **45**(5) 343–372.

Reed, R. J. 1960: Principal frontal zones of the Northern Hemisphere in winter and summer. *Bull. Amer. Met. Soc.* **41** 591–598.

Romanov, Yu. A. 1965: Mean resultant wind divergence and vorticity charts for the Indian Ocean. *Trudy*

---

* A complete list of earlier references can be found in the paper by Yoshino (1969b).

Gos. Okeanogr. Inst., Acad. Nauk SSSR (78) 119–127.

Romanova, N. A., Romanov, Yu. A. 1965: On the influence of the grid steps on the wind speed divergence. *Trudy Gos. Okeanogr. Inst.*, Acad. Nauk SSSR (78) 111–118.

Sadler, J. C. 1963: TIROS observations of the summer circulation and weather patterns of the eastern North Pacific. *Symp. Tropical Meteorology* 553–571.

Sadler, J. C. 1969: Average cloudiness in the tropics from satellite observation. Honolulu, 23p.

Sandoval, A. 1967: Background studies for a climatology of the intertropical convergence zone in the western and central Pacific area. The Univ. of Wisconsin, Ph.D. Thesis, 127p.

Thompson, B. W. 1951: An essay on the general circulation of the atmosphere over Southeast Asia and the West Pacific. *Q. Jour. Roy. Met. Soc.* 77 569–597.

Wagner, A. 1931: Zur Aerologie des indischen Monsuns. *Gerl. Beitr. Geophys.* 30 196–238.

Wiederanders, C. J. 1961: Analyses of monthly mean resultant winds for standard pressure levels over the Pacific. *Met. Div., Hawaii Inst. Geophys., Univ. Hawaii, Sci. Rep.* (3) Contract No. AF19 (604)–7229, 1–83.

Wyrtki, K. 1966: Seasonal variation of heat exchange and surface temperature in the North Pacific Ocean. Hawaii Institute Geophys., Univ. Hawaii, HIG-66-3, 8p.+72figs.

Yoshimura, M. 1967: Annual change in frontal zones in the Northern Hemisphere. *Geogr. Rev. Japan* 40 393–408 (in Japanese with English abstract).

Yoshimura, M. 1970: A climatological study on the ITCZ. *Tenki* 17(3) 119–126 (in Japanese).

Yoshino, M. M. 1963: Rainfall, frontal zones and jet streams in early summer over East Asia. *Bonner Met. Abhandl.* (3) 1–127.

Yoshino, M. M. 1965: Four stages of the rainy season in early summer over East Asia (Pt. I). *J. Met. Soc. Japan II*, 43 231–245.

Yoshino, M. M. 1967a: Maps of the occurrence frequencies of fronts in the rainy season in early summer over East Asia. *Sci. Rep. Tokyo Kyoiku Daigaku, Sec. C* 9 (89) 211–245.

Yoshino, M. M. 1967b: Atmospheric circulation over the Northwest Pacific in summer. *Met. Rdsch.* 20 45–52.

Yoshino, M. M. 1969a: Intertropical convergence zone and polar frontal zone over South, Southeast and East Asia: a climatological view. *Ann. Met. N. F.* (4) 212–220.

Yoshino, M. M. 1969b: Climatological studies on the polar frontal zones and the intertropical convergence zones over South, Southeast and East Asia. *Climatol. Notes, Hosei Univ.* (1) 1–71.

Data Sources

Data 1: Meteorological Department, Thailand 1965: Upper winds over Southeast Asia and neighbouring areas (2,000 ft).

Data 2: Meteorological Department, Thailand 1965: Upper winds over Southeast Asia and neighbouring areas (5,000 to 40,000 ft).

Data 3: Joint Task Force Seven: Mean monthly upper tropospheric circulation over the tropical Pacific during 1954–1959.

Data 4: Marine Division, Japan Meteorological Agency, Tokyo: Marine climatological tables of the North Pacific Ocean, 1942–1960 (Pt. 1), 1961, 1962, and 1963.

PART III

SUMMER MONSOON AND DISTRIBUTION OF WETNESS

CHARACTERISTIC FEATURES OF GENERAL CIRCULATION IN THE ATMOSPHERE AND THEIR RELATION TO THE ANOMALIES OF SUMMER PRECIPITATION IN MONSOON ASIA

Hideo WADA

Abstract: The monthly mean height, temperature, and wind fields and their monthly changes over Monsoon Asia during the summer monsoon season were studied from a statistical and climatological viewpoint, using normal monthly mean data. Some new features of the high cells called the Ogasawara, Pacific, and Tibetan Highs were pointed out through analysis of data for both the troposphere and the stratosphere. Relationships between summer precipitation in Monsoon Asia and large-scale circulation patterns at the 500 mb level over the Eastern Hemisphere were studied from a synoptic viewpoint, making use of correlation field maps and composite maps. It was clarified synoptically that circulation patterns not only at the 500 mb level but also at the 100 mb level greatly influence summer precipitation in Monsoon Asia.

1. INTRODUCTION

In recent years, owing to the considerable increase in upper-air data, many investigations have been made of the characteristics of general circulation and summer precipitation in Monsoon Asia (e.g. Ramanathan, 1960; Flohn, 1957, 1960). The studies made so far, however, generally have not used data covering conditions in the troposphere over Monsoon Asia over a long period of time. Neither the cause of the formation of summertime anticyclones over the Tibetan Plateau nor the relationship between summer precipitation and the circulation pattern in Monsoon Asia has yet been clarified sufficiently. To answer these questions, one can first study the basic features of the general circulation over Monsoon Asia, using long-term climatological data for the troposphere and if possible also for the stratosphere, and then tackle each problem from the physical viewpoint, using more detailed data covering a number of years.

In this paper, the seasonal changes in the general circulation over Monsoon Asia will be examined from the climatological viewpoint and many previously noted characteristics of summer circulation will be reaffirmed statistically, while some new features will also be pointed out. Furthermore, a study will be made of the relationship between the summer precipitation for several selected representative stations in Monsoon Asia and the circulation pattern in the middle and upper troposphere over the Eastern Hemisphere.

The basic materials used are the seven-level monthly mean normal charts over the Northern Hemisphere for the period 1951–1960, published by the Free University of Berlin (1969),

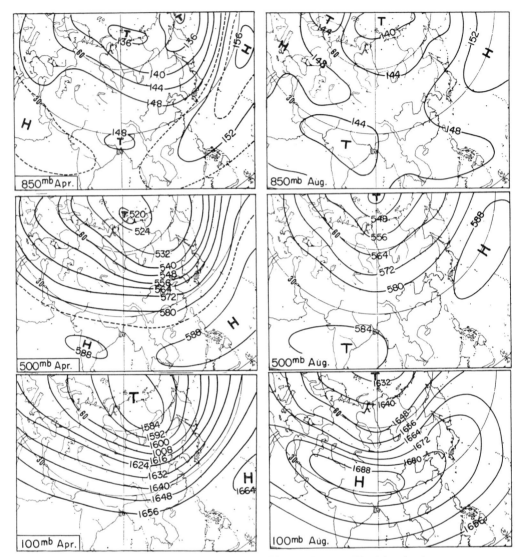

Fig. 1. Monthly mean normal contour patterns at 850, 500, and 100 mb levels for April (decameters).

Fig. 2. Same as Fig. 1 but for August.

which are the most reliable charts available at present. The grid-point data are read from the charts and the monthly changes in height, temperature (thickness), and wind field maps are calculated from these. The monthly mean 500 mb data at grid points over the Northern Hemisphere for the period 1947–1968, which are used to compute the correlation coefficients, were prepared by the Japan Meteorological Agency.

2.   SOME FEATURES OF THE HIGH CELLS OVER MONSOON ASIA, INCLUDING SEASONAL CHANGES IN HEIGHT PATTERN

Seasonal changes in the monthly mean normal height pattern were studied at seven levels: Surface, 850, 700, 500, 300, 200, and 100 mb. Two examples of monthly mean charts for April and August are shown in Figs. 1 and 2, with contour patterns for only three levels; the plottings of the observed raw data are omitted.

Seasonal changes in height patterns from winter to summer can be summarized as follows:

1)   The lower troposphere (850 mb): A large-scale heat low is formed over India in May, causing the onset of the SW monsoon over South Asia. The Pacific High shifts its center northward with the passing of the season and spreads to the east and west in August, with its center located around the Hawaiian Islands.

2)   The middle troposphere (500 mb): The high cell situated over the Philippine Islands in April moves its axis northward as the season advances, its center moving northeastward. A large-scale low is formed over India in June, and in general the SW monsoon prevails over South Asia during June–September. It should be noted here that the mean trough situated over the Bay of Bengal in May gradually moves westward and coincides with the mean trough in the middle latitudes; this phenomenon was named the "extended trough" by Cressman (1948).

3)   The upper troposphere (100 mb): The subtropical high gradually shifts its axis northward as the season advances. An extensive high forms over the Tibetan Plateau in June and persists during June–September, while reaching its maximum intensity. This period also coincides with the time of the SW monsoon over Monsoon Asia.

An interesting feature of the seasonal change in the positions of the high cells in the troposphere is shown in Fig. 3, where the tracks of the centers of subtropical high cells on the monthly mean normal charts at 850, 500, and 100 mb over Monsoon Asia are depicted for the period from January to September. The lower tropospheric high center (850 mb) is located north of the Philippine Islands in January, after which it moves steadily eastward, reaching the middle of the Pacific Ocean in summer. The middle tropospheric high center (500 mb) is situated somewhat east of the Philippine Islands in winter and gradually migrates north to northeastward, until it is far southeast of Japan in summer. On the other hand, the upper tropospheric high center (100 mb), which is located far southeast of the Philippine Islands in winter, abruptly moves to India between April and May and persists over the Tibetan Plateau during the monsoon season. The daily map series indicates that the abrupt change in position of the 100 mb high cell seems to occur generally through the westward

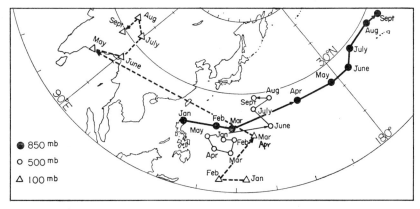

Fig. 3.   Tracks of the subtropical high centers on the monthly mean normal charts at 850, 500, and 100 mb levels from January to September over Monsoon Asia.

migration of the high cell at the 100 mb level, which extends from the Philippines westward, and rarely through the eastward migration of the high cell over Africa. This abrupt change in position of the high cell at the 100 mb level in May, coinciding as it does with the appearance of a closed surface low over northern India, suggests a possible mechanism for the onset of the summer monsoon over South Asia.

The southward retreat of the high cells at each level after the summer monsoon is roughly a reversal of the course followed from winter to summer.

An interesting feature of the tropospheric high cells, which are located on the far eastern side of the Philippine Islands in wintertime, is that they migrate with the passing of the season in directions quite different from each other. To confirm the existence of three individual cells in the atmosphere, the monthly mean normal height profiles at seven levels along the 30°N latitude circle for July are given in Fig. 4. From this figure, it is apparent that there are three highs in the troposphere over Monsoon Asia: A lower tropospheric high at 850 mb, called the Pacific High, which is formed by the effect of low sea-surface temperature and is located in the middle of the Pacific Ocean (170°W); a middle tropospheric high, called the Ogasawara High, which is noticeable at 700–500 mb levels and appears towards the southern part of Japan (150°E); and an extensive high at 300–100 mb levels, called the

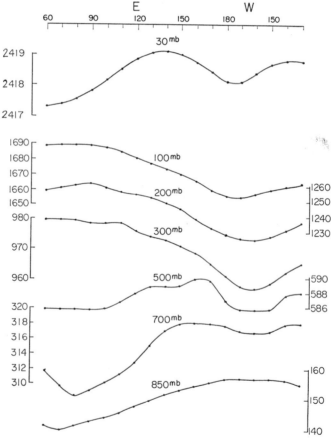

Fig. 4. Monthly mean normal height profiles of seven levels along the 30°N latitude circle from 60°E to 130°W for July (decameters).

Tibetan High, which is situated over the Tibetan Plateau (80°E). In addition, the profile for the stratosphere (30 mb), which was constructed using mean data for the periods 1957–1960 and 1963–1967, and the 700 mb profile are quite similar, both showing a maximum height at about 140°E.

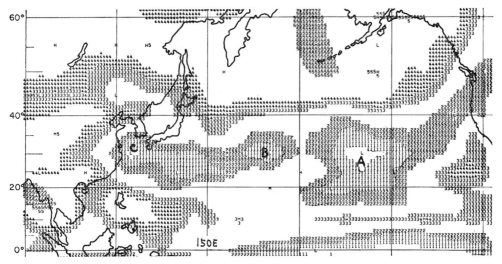

Fig. 5.   Monthly mean chart of brightness for August 1967 (Taylor and Winston, 1968).

From these facts, it is clear that the subtropical Pacific High, as it has usually been called by meteorologists, is generally composed of at least two cells, and sometimes three. This is confirmed by the monthly mean brightness chart for August 1967, as shown in Fig. 5, in which three centers of lower brightness, A, B, and C, can be seen. This chart was constructed from ESSA 3 and 5 digitized pictures (Taylor and Winston, 1968). Generally, an area of lower brightness over the ocean roughly coincides with the center of a high cell, indicating a cloudless region. Areas A and C on the brightness chart are regarded as the centers of the Pacific and Ogasawara Highs respectively, and the positions of A and C roughly coincide with the positions of the respective highs on the normal monthly mean chart above. Area B may correspond to the maximum height at 170°E in the 500 mb profile in Fig. 4.

In recent years, many studies have been made in Japan of the behavior of the anticyclone over Monsoon Asia and it has become clear that the Tibetan High greatly influences summer weather in Japan (Neyama, 1965; Asakura, 1968; Kurashima, 1968). For example, in the summer of 1967, western Japan suffered from an extremely severe drought created by the persistent high centered over the East China Sea that is seen as the low brightness area in Fig. 5. This high persisted throughout the summer under the effect of the eastward extension of the Tibetan High and the westward migration of the Ogasawara High.

The wind field generally changes in accordance with changes in the pressure pattern. Therefore, the seasonal change in wind field may be roughly understood from monthly mean contour patterns such as those in Figs. 1 and 2. It is well known that over Monsoon Asia a SW wind prevails in the lower troposphere whereas an easterly wind prevails in the upper troposphere during the summer monsoon season. As an example of the seasonal changes in winds in the upper troposphere, the annual change in observed winds at a monthly mean 100 mb level along the 80°E meridian is given in Fig. 6, together with the

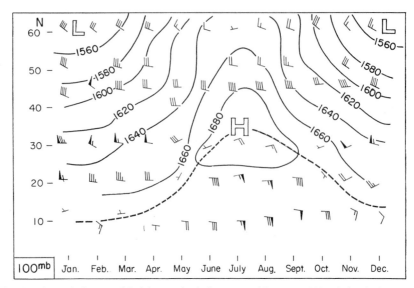

Fig. 6. Annual change of heights and winds at monthly mean 100 mb level along 80°E (decameters).

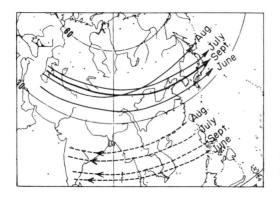

Fig. 7. Position of monthly mean westerly and easterly jet stream axes at 100 mb level for the months June–September.

contours. This figure indicates that an easterly wind prevails throughout the year at the lower latitude of 10°N. However, the easterlies gradually stretch northward with the passing of the season, reaching about 30°N latitude with a maximum wind intensity in July or August, during which time the Tibetan High over the plateau also attains maximum intensity. To illustrate this further, the monthly mean positions of the westerly and easterly jet stream axes at a 100 mb level are shown in Fig. 7 for the period June–September. The months of July and August, during which the intensity of monsoon precipitation reaches its peak in South Asia, correspond quite closely with the period when the westerly and easterly jet stream axes are displaced northward. This fact may be related to the mechanism of monsoon precipitation and also to the year-to-year variation in the amount of monsoon precipitation.

3. Seasonal Changes in the Temperature Field in the Troposphere over Monsoon Asia

In order to study the temperature distribution and its seasonal change in the troposphere, maps of monthly mean normal thickness and thickness change for the 500–850 mb and 100–500 mb levels over the Northern Hemisphere were constructed for each month. These maps illustrate an interesting feature of the upper troposphere (100–500 mb): Generally, seasonal temperature changes in the atmosphere over the Northern Hemisphere do not occur uniformly but show certain peculiarities in the upper troposphere, as seen in Figs. 8 and 9, where monthly changes of thickness from March to April and from May to June are depicted. In spite of the passage of time from March to April, there is only a small decrease in thickness over the Tibetan Plateau, in striking contrast to the marked increase in thickness over high latitudes (Fig. 8). However, a remarkable increase in thickness occurs over the Tibetan Plateau from April to May, and continues into August, as shown in Fig. 9. Meanwhile, a thickness change pattern similar to the May–June pattern persists in the monthly mean thickness distribution with an extensive, large thickness over the Tibetan Plateau during the monsoon season. A pattern similar to that for March–April (as seen in Fig. 8) cannot, however, be found in the lower troposphere (500–850 mb), where a monthly increase in thickness generally occurs toward summer over the Northern Hemisphere, with its maximum area in high latitudes. Presumably, the temporary decrease of temperature (thickness) in the upper troposphere over the Tibetan Plateau from March to April is caused by the cooling effect of snow melting on the highlands in April when the temperature in the middle troposphere rises about 0°C. It is, however, improbable that the marked increase of thickness in the upper troposphere in May is caused by the heating effect of the highlands, which are bare of vegetation. It may be due to the release of latent heat through condensation from the SW monsoon, since the increase in thickness in May roughly coincides with the onset of the SW monsoon over the region. In any case, it can be seen that the normal temperature

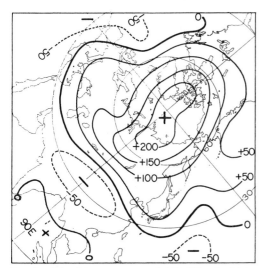

Fig. 8. Monthly mean normal 100–500 mb thickness change from March to April (meters).

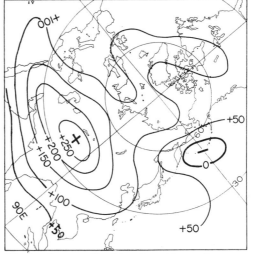

Fig. 9. Same as Fig. 8 but from May to June.

distribution and its seasonal change in the atmosphere over the Northern Hemisphere are greatly affected by the Tibetan Plateau.

Although there have been many studies of the Tibetan High (Flohn, 1957; Academia Sinica, Peking, 1958; Rangarajan, 1963), no satisfactory physical explanation of its origin has been divised to date, and the subject is still in dispute. Observation of the unique thermal feature in the upper troposphere mentioned above may make an important contribution to the study of the origin and maintenance of the Tibetan High. It is of interest to note here that the marked temperature increase in the upper troposphere over the Tibetan Plateau in May occurs in the same month as the abrupt change in position of the subtropical high at 100 mb.

To investigate temperature distribution in the subtropical high in summer in detail, the July monthly mean normal thickness profiles through the atmosphere along the 30°N latitude circle are presented for 500–850, 100–500, and 30–100 mb in Fig. 10, with height profiles for 700, 100, and 30 mb. The figure shows that the temperature throughout the troposphere is much warmer over the continent than over the ocean, but in the stratosphere there is a reverse temperature distribution showing temperature compensation. It is noteworthy that the temperature distribution of the three highs suggests quite different thermal mechanisms in vertical distribution: The Pacific High (170°W) mentioned earlier is built up in the lower troposphere by relatively cool air over the ocean, while the Ogasawara High (150°E) reaches its maximum intensity in the middle troposphere with an obscured temperature feature; the Tibetan High (80°E), which forms roughly above the large-scale depression in the lower troposphere, reaches its maximum intensity at the 200 mb level, with warm air throughout the troposphere, but with markedly cold air in the stratosphere. This relationship was also confirmed in a case study of the summertime anticyclone over South Asia by Mason and Anderson (1963).

The marked temperature compensation between the troposphere and stratosphere over the monsoon depression around India is quite an interesting subject for us, because it is well known that there is marked compensation between the Aleutian Low in the troposphere and the Aleutian High in the stratosphere, although the vertical temperature distribution pattern of the Tibetan High is the opposite of that over the Aleutian. It is not valid to apply the higher latitudes concept directly to the lower latitudes, but it is worthwhile to study large-scale circulation patterns lasting a few months, such as the Tibetan High and the Aleutian High, through comparison of the energy cycle in the atmosphere of each. This problem will be reserved for future study.

Very recently, it has become clear that not only the large-scale circulation pattern in the troposphere but also the pattern in the stratosphere greatly affects the *Grosswetter* over the Northern Hemisphere during the winter season (Wada and Asakura, 1967). Applying this concept to the summer season, the relationship between the height and thickness in the troposphere and in the stratosphere was studied for the summer seasons of 1963 and 1964, through calculation of the correlation coefficients among five-day mean 500, 100, and 30 mb heights and also among five-day mean 500–1,000, 100–500, and 30–100 mb thicknesses over the Northern Hemisphere. Since calculations of correlation coefficients were made for 36 cases of five-day mean values, the level of significance is 0.32 by the 5% criterion. The results for the Monsoon Asia area are summarized as follows:

1) Height relationship between 500 and 100 mb: high significant positive correlation areas are found over Lake Balkhash, far north of the Tibetan Plateau, and over the East China Sea, and a small negative area over Indochina and the Philippine Islands. No significant correlation is found over India.

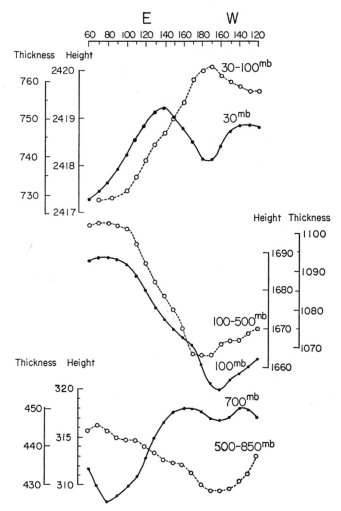

Fig. 10. Monthly mean normal height and thickness profiles in the troposphere and stratosphere along the 30°N latitude circle from 60°E to 120°W for July (decameters).

2) Height relationship between 500 and 30 mb: for investigation of height relationships over the Ogasawara High, see the example of height correlation field over Monsoon Asia illustrated in Fig. 11. Note that there is no significant relation between 500 and 30 mb over Monsoon Asia except over the East China Sea, which indicates that the lower pressure at 500 mb level over India has no relation to the height in the stratosphere. There is a significant positive correlation area over the East China Sea, which is also seen from the normal profiles of 700 or 500, and 30 mb in Figs. 4 and 10.

3) Height relationship between 100 and 30 mb: generally there is a high positive correlation area over Japan, extending east and west, and a small positive area over the Tibetan Plateau, indicating that the Tibetan High has little relation to the height in the stratosphere.

4) Thickness relationship between 500–1,000 and 30–100 mb: a negative correlation area is generally seen over Monsoon Asia, with a marked negative region over Lake Bal-

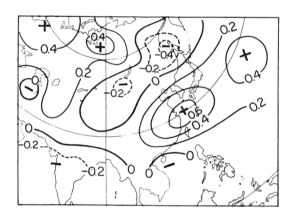

Fig. 11. Distribution of correlation coefficients of five-day man heights between 500 and 30 mb for two summers (June–August) of 1963 and 1964.

khash (−0.6). This fact means that temperature compensation between the troposphere and stratosphere in the summer season takes place mainly over Monsoon Asia.

As stated earlier, the Ogasawara High, which exists in the middle troposphere in the summer season, sometimes migrates westward and amalgamates into the Tibetan High at its eastern edge, bringing a severe drought to western Japan. From the significant positive correlation of the height over the East China Sea between 500 and 30 mb, it can be hypothesized that the decay or development of the Ogasawara High may be affected by the degree of circulation in the stratosphere. It should be noted here that data for the 500 mb level are not sufficient to support hypotheses about the behavior of the upper tropospheric high, at least over India, because there is no relation between the 500 and 100 mb heights over India, as is also indicated by the vertical distribution of monthly mean height profiles shown in Fig. 4.

From the foregoing statements, it is statistically clear that the Pacific, Ogasawara, and Tibetan Highs over Monsoon Asia differ greatly in the structure of their height and temperature fields. These differences may play an important role in determining the summer weather in Monsoon Asia.

4. SOME FEATURES OF SUMMER PRECIPITATION IN MONSOON ASIA

In studying the general features of summer precipitation in Monsoon Asia, data from eight selected stations were used, as shown in Fig. 12; the eight stations were Bombay, Calcutta, Madras, Bangkok, Saigon, Hong Kong, Taipei, and Kagoshima. The data used are the monthly precipitation amounts at these stations and the monthly mean 500 mb height values at each grid point over the Eastern Hemisphere during the period June–September in the years 1947–1968.

Since the amount of precipitation itself is generally localized, as is well known, it is impossible to demonstrate an exact correlation between precipitation amount in each area, especially in such a vast region as Monsoon Asia. Nevertheless, simultaneous correlation coefficients of the monthly precipitation amounts at each station were calculated tentatively for the four months from June to September. Sample correlation coefficient values for July are shown in Table I. The values were obtained from data covering a period of 22 years, and the significant level of the correlation coefficient is 0.42 by the 5 % criterion.

It can be seen from Table I that there is a significant relationship in precipitation amount

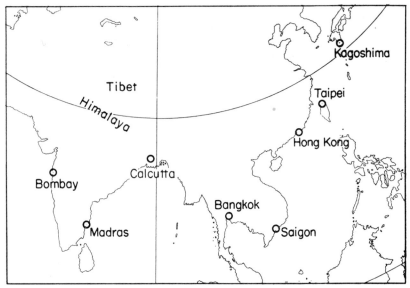

Fig. 12. Map of selected stations in Monsoon Asia.

Table I. Simultaneous correlation coefficients of monthly precipitation amounts among eight stations for July.

July	1	2	3	4	5	6	7	8
1. Bombay	1.00	0.52	0.17	0.18	−0.12	0.09	0.22	−0.74
2. Madras		1.00	0.10	0.27	0.10	0.01	0.26	−0.49
3. Bangkok			1.00	0.17	−0.10	−0.07	0.41	−0.34
4. Saigon				1.00	−0.09	−0.34	0.15	−0.20
5. Hong Kong					1.00	0.27	−0.09	−0.19
6. Taipei						1.00	−0.35	−0.11
7. Kagoshima							1.00	−0.39
8. Calcutta								1.00

between Calcutta and two out of the other seven stations: Bombay and Madras. The inverse relationship between the monthly precipitation amount at Calcutta and those at Bombay and Madras, which was also confirmed by Subbaramayya (1968), holds throughout the June–August period; the negative relationship turns into a positive one in September. It is interesting to note that the monthly precipitation amount in July for Calcutta has a negative correlation with that of all other stations, although in comparison to some of the stations the coefficients are very small. This suggests that the summer precipitation in Monsoon Asia may be affected by some factor that controles summer precipitation in Calcutta.

From the results of all calculations, significant coefficients have been selected and are shown in Table II. It is difficult to explain from a synoptic viewpoint the interrelations of the monthly precipitation amounts mentioned above, but it is noteworthy that the block areas with similar amounts of precipitation for June–August can be estimated from the significant values in Table II. However, for September, the amount of monthly precipitation in Hong Kong and Taipei shows a strong negative correlation, despite extensive calculations; this may be caused by the frequent passage of typhoons over Taipei during the month.

Table II.   Significant correlation coefficient values of monthly precipitation
amounts among eight stations for June–September.

June	July	August
Bombay-Taipei −0.47	Bombay-Madras 0.52	Bombay-Calcutta −0.40
Bombay-Calcutta −0.45	Bombay-Calcutta −.74	Hong Kong-Kagoshima −0.56
Bangkok-Calcutta 0.44	Madras-Calcutta −0.49	Hong Kong-Calcutta 0.56
Hong Kong-Taipei 0.63	Bangkok-Kagoshima 0.41	Kagoshima-Calcutta −0.55

September
Bangkok-Kagoshima 0.44
Hong Kong-Taipei −0.52

The relationship of circulation pattern to summer precipitation in Monsoon Asia, especially in India, has been studied extensively (e.g. Pisharoty and Asnani, 1960; Ramdas, 1960). In order to check the relationship, the correlation coefficients between the monthly precipitation amount at each station and the zonal index were calculated for each month during the period June-September. In this calculation, a cube-root transformation of the monthly precipitation amount was made to normalize its frequency distribution. The monthly mean zonal index is usually calculated from the difference in height between two latitudes on the monthly mean 500 mb chart, and here two zonal indices for 40–60°N and 30–50°N are calculated for the Eastern Hemisphere. In order to study the circulation pattern over Monsoon Asia, the index should be calculated for more lower latitudes, but the paucity of long-term data in lower latitudes precluded this. Only five significant coefficients were obtained: At 40–60°N, Bangkok −0.42 for July, Hong Kong −0.50 for August, and Kagoshima +0.42 for September; at 30–50°N, Kagoshima −0.52 for July and Calcutta +0.46 for August. Although it cannot be concluded that there is always a good correlation between the monthly precipitation amount in Monsoon Asia and the zonal index for the Eastern Hemisphere, the zonal index of 30–50°N seems to be more useful than that of 40–60°N in studying their relation in India.

To study the relation between the monthly precipitation amount and the circulation pattern over Monsoon Asia in more detail, correlation field maps for July and August were constructed for all of the above-mentioned stations, through calculation of the correlation coefficients between the monthly precipitation amount at each station and the monthly mean 500 mb height grid-point values over the Eastern Hemisphere, with data covering 22 years. The results for both months show that there are significant correlations at each station except Taipei. For example, the longitudinal distributions of correlation coefficients along the 40°N circle at Bangkok, Saigon, and Hong Kong are shown in Fig. 13, which was drawn using the values extracted from the correlation field maps. The figure indicates that significant negative regions exist in the middle of Siberia. On the other hand, a climatological mean chart of height for the summer season shows the existence of an extensive trough in the middle of Siberia, which extends from the surface up to the mid-troposphere, as was pointed out previously. With the existence of the trough, and on the basis of Fig. 13, it can be said that as the climatological trough in the middle of Siberia intensifies, the amount of precipitation at Bangkok, Saigon, and Hong Kong increases.

It is clear from the facts mentioned thus far that precipitation in Monsoon Asia should be understood not as a local phenomenon but as an associated one, affected by circulation over the Eastern Hemisphere. However, since it is difficult to describe in detail the different relations between the monthly precipitation amount for each of the eight stations in Monsoon Asia and the circulation pattern over the Eastern Hemisphere, further discussion in this paper will focus on the summer precipitation at three stations in India.

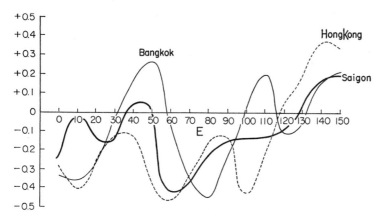

Fig. 13. Longitudinal distribution along the 40°N latitude circle of correlation coefficients between the cube-root of monthly precipitation amounts in Bangkok, Saigon, and Hong Kong, and monthly mean 500 mb heights for July (1947–1968).

5. THE RELATIONSHIP BETWEEN SUMMER MONSOON PRECIPITATION IN INDIA AND THE CIRCULATION PATTERN OVER THE EASTERN HEMISPHERE

Although many investigations have been made of the onset of the summer monsoon or of precipitation in India since the original investigation by Walker, forecasting monsoon precipitation in India remains problematical (Charles, 1953). In order to make a forecast of monthly precipitation, it is necessary first to study the relationship between the monthly precipitation amount and the circulation pattern, and then to clarify the relationship from a physical and synoptic viewpoint, as pointed out by Stidd (1954).

Recently, for example, it was found that there was a close relation in winter between the monthly precipitation amount on the Japan Sea side of Japan and the circulation pattern over the Far East; this conclusion was based on a study using correlation field and composite maps of the monthly mean 500 mb height over the Northern Hemisphere. Furthermore, the patterns of simultaneous and lag correlations were explained synoptically and have been utilized for practical long-range forecasting of precipitation, the technique being called a synoptic-correlation method. On the other hand, in Japan many studies have been made of the relation between Baiu rains affected by anticyclonic conditions over the Sea of Okhotsk and the SW monsoon over India, and also of the mechanism of the Okhotsk High and the Tibetan High (Suda and Asakura, 1955; Murakami, 1957, 1958; Asakura, 1968). It is known that year-to-year variations in summer weather in Japan have a close connection with the development of the summer monsoon over India. Therefore, it is also very important for us to understand the summer monsoon over India.

In order to study summer monsoon precipitation in India from the climatological viewpoint, it is first necessary to know in detail the features of seasonal change in the circulation pattern over India. Figures 14 and 15 show the normal monthly mean zonal 500 mb height profiles along the 30°N and 50°N latitude circles for the periods April–September and May–

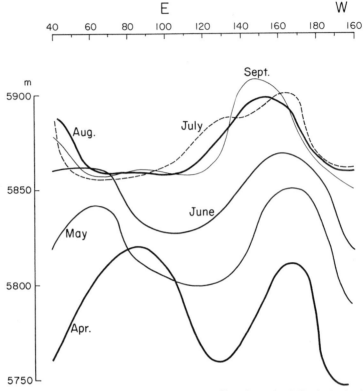

Fig. 14. Monthly mean normal 500 mb height profiles along the 30°N latitude circle for April–September.

September respectively. From Fig. 14, it can be seen that in the lower latitude the trough is present at about 130°E in April but retrogrades gradually westward as the season passes, until it reaches the area between 60°E and 110°E, where it persists during the July–September period. The profiles in the higher latitude at 50°N in Fig. 15 show that a deep anchor trough persists in the middle of the Pacific Ocean during the period of May–September, forming a secondary trough at about 90°E in July and August. On the basis of these figures, it should be emphasized that the circulation patterns in the middle of the troposphere expressed by 500 mb height do not show the same features throughout the monsoon season, and that from a climatological viewpoint the variation of monsoon situations is caused by the change in monthly circulation patterns in the mid-troposphere. However, since there is a common feature in the patterns for July and August, the period of maximum development of the SW monsoon over India, further attention in this paper will be centered mainly on the monthly precipitation amount for these months.

Many studies have been made of the SW monsoon over India, and recently Chang (1967) and Lockwood (1965) have reviewed this subject. Ramaswamy (1965) has shown synoptically that there is a close relation between the precipitation in India and the zonal index. On the other hand, in my research on the relation between the monthly precipitation amount in India and the zonal index, a significant relation was found only for Calcutta in August, as was pointed out earlier. A scatter diagram of this relationship for each year in Calcutta is given in Fig. 16: The black circles show plots for July and the white ones for August; the

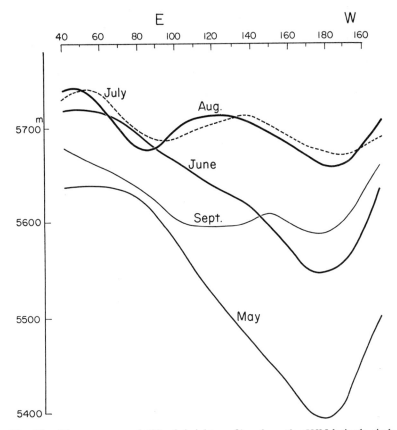

Fig. 15. Monthly mean normal 500 mb height profiles along the 50°N latitude circle for May–September.

monthly precipitation amount (given as the cube-root) is the ordinate and the monthly mean zonal index anomaly over the Eastern Hemisphere is the abscissa. The figure suggests that a high index pattern is favorable to high precipitation in Calcutta, with a positive correlation between these features. The figure illustrates, however, that there were two exceptional years, 1952 and 1961. Mothe and Wright (1969) suggested in their study on the onset of the Indian SW monsoon that in some years, e.g. in 1961 when the monsoon was very severe, the behavior of the middle latitude 500 mb flow apparently played only a small part in the mechanism of onset of the monsoon. The results of their study indicated that the middle latitude flow does not always control monsoon precipitation in Calcutta, and that there may be another factor involved. The value of the correlation coefficients for July was first calculated as $+0.25$, but this was raised to $+0.61$ in later calculations which excluded data for 1952 and 1961. The data mentioned above confirms statistically that the monthly precipitation amount in July or August in Calcutta generally has a positive correlation to the zonal index defined by the height difference between 30°N and 50°N on the monthly mean 500 mb chart.

Since there are significant negative correlations between the monthly precipitation amount in Calcutta and those in Madras and Bombay, the precipitation in India will be described on the basis of the precipitation amount in Calcutta, although this description remains tenta-

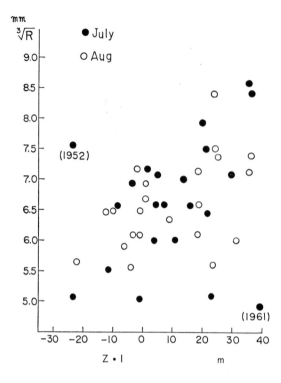

Fig. 16. Scatter diagram of the cube-root of monthly precipitation amounts in Calcutta and the zonal index anomaly (30–50°N) on the monthly mean 500 mb chart over the Eastern Hemisphere for July and August.

tive. From the field maps giving simultaneous correlations between the monthly precipitation amount in Calcutta and the monthly mean 500 mb height at every grid-point over the Eastern Hemisphere for the period June–September, one example, for August, is shown in Fig. 17, where two significant negative regions are found around Lake Baikal and Turkey. Generally speaking, a correlation field pattern is considered analogous to a height anomaly one; therefore, the correlation field pattern in Fig. 17 proves incontestably that the so-called blocking situation in higher latitudes brings about heavy precipitation in Calcutta, accompanied by an invasion of cold air masses into the two regions of Lake Baikal and Turkey. The patterns for Bangkok, Saigon, and Hong Kong are almost identical to this one, except for some differences over the Tibetan Plateau. Returning to Calcutta, it is interesting to note that a positive correlation area is located around the Tibetan Plateau. This suggests that the circulation over the Tibetan Plateau also has a considerable influence on precipitation in Calcutta, in view of the fact that a positive correlation holds between 500 and 100 mb heights over the Tibetan Plateau, as described above.

On the basis of Fig. 17, it should be emphasized that the monthly precipitation amount for Calcutta bears a close relation to the circulation pattern over Europe and Asia, and that the cold air masses centered around Lake Baikal and Turkey are favorable to heavy precipitation in Calcutta; there are also indications in the figure of a low index pattern in higher latitudes and a high one in lower latitudes. On the other hand, it has become clear recently that the polar vortex in the middle troposphere influences the *Grosswetter* in middle latitudes in wintertime (Wada and Asakura, 1967). Thus, it should be stressed here that the summer precipitation in Monsoon Asia must also be closely associated with the behavior of the polar vortex at higher latitudes, which causes a blocking situation over the Eastern Hemisphere. Sutcliffe and Bannon (1954) have already demonstrated that summer monsoon precipitation in India is greatly affected by a blocking situation over Europe. This effect was reaffirmed more precisely in the present study, through the presentation of correlation field maps.

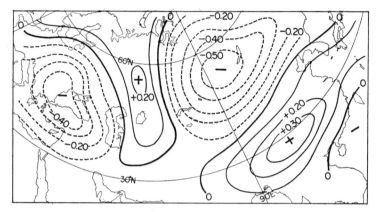

Fig. 17. Simultaneous correlation field between the cube-root of monthly precipitation amounts in Calcutta and the monthly mean 500 mb height for August over the Eastern Hemisphere (1947–1968).

The conditions described above for August are also seen in June and July, but with smaller correlation coefficients. However, the features of the correlation field map for September (not shown here) are almost the opposite of those for August in Fig. 17, two significant negative areas having changed into significant positive ones. It is possible that, in September, a deep meandering of westerly flow with an intensification of the trough at about 10°E brings about heavy precipitation in Calcutta.

Composite maps of the monthly mean 500 mb height anomaly have been constructed for three cases of extremely heavy and light monthly precipitation amounts in Calcutta, selected from the data for the period 1947–1968. An examination of these maps shows a common feature: Westward displacement and deepening of the climatological trough at about 90°E are quite favorable to heavy precipitation in Calcutta, being conditions that bring a stronger SW wind over North India through the monsoon season. In addition, coincidence of the longitudinal positions of the trough at both the higher and lower latitudes is also favorable to heavy precipitation in Calcutta. Such conditions may at the same time, however, favor scanty precipitation in Bombay or Madras, perhaps causing the SW monsoon to break in these areas, as suggested by Ramaswamy (1962). It is clear from these results, although all the correlation field maps and composite maps have not been presented here, that the peculiarities of the circulation pattern in the middle troposphere over the Eastern Hemisphere have a considerable effect on summer monsoon precipitation in India.

As mentioned previously, summer precipitation in western Japan is greatly affected by the behavior of the Tibetan High. It has been demonstrated that the SW monsoon in India is also affected by the Tibetan High, with the development of an easterly jet stream along the southern part of India (Koteswaram, 1958; Mason and Anderson, 1963). In order to confirm the relationship between the Tibetan High and precipitation in India, the months of July of 1964 and 1968 were selected as examples of extremely wet months and July of 1966 as an extremely dry one for Calcutta in recent years. To find some common feature between these two extremes in the circulation pattern of the upper troposphere, two maps of height differences of the monthly mean 100 mb for wet and dry months were constructed, as shown in Fig. 18. Note that the patterns are quite similar. This shows that the precipitation in Calcutta is affected by the decay or development of the Tibetan High: When the Tibetan High is strengthened and extends southward, it causes a dry month in Calcutta but brings much precipitation to Madras and Bombay. The same feature was found in the extremely wet month of August in Calcutta in 1968.

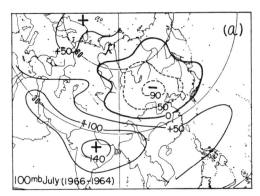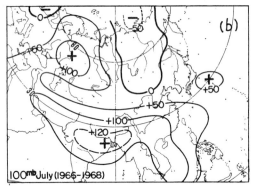

Fig. 18.   a: Height difference of the monthly mean 100 mb for dry (1966) and wet (1964) Julys in Calcutta (meters). b: Height difference of the monthly mean 100 mb for dry (1966) and wet (1968) Julys.

Thus, it should be stressed that the Tibetan High plays an important role in the amount of summer monsoon precipitation in India; it is obviously important to know the behavior of the Tibetan High, for forecasting precipitation in Monsoon Asia, because the upper tropospheric pattern over there is persistent or moves quite slowly.

7. CONCLUSIONS

The results of this investigation, which was based on monthly mean 500 mb data for 22 years and the normal data for seven levels for the period 1951–1960 over the Northern Hemisphere, can be summarized as follows:

1)   The subtropical high over the Pacific Ocean is generally composed of two or more cells in the summertime, which differ in both vertical height and temperature distribution.

2)   The Tibetan High, which is most noticeable at the 200 mb level, with warm air throughout the troposphere and cold air in the stratosphere, greatly affects the *Grosswetter*, not only over South Asia but also over western Japan.

3)   Seasonal changes in temperature in the troposphere do not occur uniformly over the Northern Hemisphere; from March to April, a small decrease in temperature appears in the upper troposphere over the Tibetan Plateau, while a marked increase in temperature occurs over the higher latitudes. After May, an intense increase in temperature in the upper troposphere continues until July over the Tibetan Plateau.

4)   Generally, the Dines' compensation theory concerning troposphere and stratosphere holds true over the Northern Hemisphere during the summer season.

5)   In summer, there is a significant positive correlation between 500 and 30 mb heights over the Ogasawara High, which often persists over southern Japan and causes a severe

drought in western Japan. On the other hand, there is no significant relation between 500 and 100 mb heights over India. Therefore, it is useless to discuss the behavior of the Tibetan High over India at the 100 mb level by using the 500 mb height data.

6)   The monthly precipitations for July and August in Calcutta, Bangkok, Saigon, and Hong Kong are greatly affected by the behavior of the trough over Siberia in the same months, a significant negative correlation being shown between the 500 mb height over Siberia and monthly precipitation amounts at these stations.

7)   The monthly precipitation amounts at stations in Monsoon Asia do not always correlate with the zonal index for the Eastern Hemisphere. The monthly precipitations for July and August in Calcutta have significant positive correlations to the zonal indices expressed by the height difference between the 30°N and 50°N latitude circles on the monthly mean 500 mb level over the Eastern Hemisphere.

8)   In July and August, the monsoon precipitation in Calcutta is greatly affected by the occurrence of a blocking pattern in the mid-troposphere over the Eastern Hemisphere, which was demonstrated by the significant negative correlations between the 500 mb height around Lake Baikal and around Turkey, and the amount of precipitation in Calcutta. However, in September, this correlation reverses itself.

9)   The intensification of the Tibetan High in July and August generally causes light precipitation in Calcutta and heavy precipitation in Madras and Bombay.

This research was conducted for the purpose of investigating the basic problem of summer precipitation in Monsoon Asia and analyzing it from the viewpoint of synoptic meteorology. Though many questions remain unanswered, as pointed out in this paper, it is hoped that the results presented here may shed some light upon the further study of long-range forecasting and climatology of precipitation in Monsoon Asia.

References

Academia Sinica, Peking 1958: On the general circulation over Eastern Asia (2). *Tellus* **10** 58–75.

Asakura, T. 1968: Dynamic climatology of atmospheric circulation over East Asia centered in Japan. *Pap. Met. Geophys.* **14** 1–68.

Chang, Jen-Hu. 1967: The Indian summer monsoon. *Geogr. Rev.* **57** 373–396.

Charles, N. 1953: Monsoon seasonal forecasting. *Quart. J. Roy. Met. Soc.* **79** 463–473.

Cressman, G. P. 1948: Studies of upper-air conditions in low latitudes, part II: Relations between high and low-latitude circulation. *Univ. of Chicago Dept. of Meteorol. Misc. Rept.* **24** 68–100.

Dao, Shin-yen, Chen, Lung-shun 1957: The structure of general circulation over the continent of Asia in summer. *75th Ann. Vol. J. Met. Soc. Japan* 215–229.

Flohn, H. 1957: Large-scale aspects of the summer monsoon, South and East Asia. *75th Ann. Vol. J. Met. Soc. Japan* 180–186.

Flohn, H. 1960: Recent investigations on the mechanism of the "summer monsoon" of Southern and Eastern Asia, *In* Symp. on Monsoons of the World 75–88.

Koteswaram, P. 1958: The easterly jet stream in the tropics. *Tellus* **10** 43–57.

Kurashima, A. 1968: Studies on the winter and summer monsoons in East Asia based on dynamic concept. *Geophys. Mag.* **34** 145–235.

Lockwood, J. G. 1965: The Indian monsoon—A review. *Weather* **20** 2–8.

Mason, R. B., Anderson, C. E. 1963: The development and decay of the 100 mb summertime anticyclone over Southern Asia. *Mon. Wea. Rev.* **91** 3–12.

Mothe, P. D. de la, Wright, P. B. 1969: The onset of the Indian southwest monsoon and extratropical 500 mb trough and ridge patterns over Europe and Asia. *Met. Mag.* **98** 145–155.

Murakami, T. 1957: On the seasonal variation of mean vertical velocity and atmospheric heat source over the Far East from spring to summer. *Pap. Met. Geophys.* **7** 358–376.

Murakami, T. 1958: The sudden change of upper westerlies near the Tibetan Plateau at the beginning of the summer season. *J. Met. Soc. Japan* **36** 239–247.

Neyama, Y. 1965: A relationship between the 100 mb anticyclone over Asia and the Ogasawara anticyclone. *J. Met. Soc. Japan* **43** 284–289.

Pisharoty, P. R., Asnani, G. C. 1960: Flow pattern over India and neighborhood at 500 mb during monsoon, *In* Symp. on Monsoons of the World 112–117.

Ramanathan, K. R. 1960: Monsoons and general circulation of the atmosphere—A review. *In* Symp. on Monsoons of the World 53–64.

Ramaswamy, C. 1962: Breaks in the Indian summer monsoon as a phenomenon of interaction between the easterly and sub-tropical westerly jet streams. *Tellus* **14** 337–349.

Ramaswamy, C. 1965: On the synoptic method of forecasting the vagaries of southwest monsoon over India and neighboring countries. The symposium on Meteorological results of the Indian Ocean Expedition.

Ramdas, L. A. 1960: The establishment, fluctuation and retreat of the southwest monsoon of India. *In* Symp. on Monsoons of the World 251–256.

Rangarajan, S. 1963: Thermal effects of the Tibetan Plateau during Asian monsoon. *Aust. Met. Mag.* **42** 24–34.

Scherhag, R., Mitarbeitern 1969: Klimatologische Karten der Nordhemisphäre. *Met. Abh.* **100** (1).

Stidd, C. K. 1954: The use of correlation fields in relating precipitation to circulation. *J. Met.* **11** 202–213.

Subbaramayya, I. 1968: The inter-relations of monsoon rainfall in different sub-divisions of India. *J. Met. Soc. Japan.* **46** 77–85.

Suda, K., Asakura, T. 1955: A study on the unusual Baiu season in 1954 by means of Northern Hemisphere upper air mean charts. *J. Met. Soc. Japan* **33** 233–244.

Sutcliffe, R. C., Bannon, J. K. 1954: Seasonal change in the upper air conditions in the Mediterranean-Middle East area. *Scientific Proceedings I.U.G.G. Association of Meteorology, Rome* 322–334.

Taylor, R., Winston, J. S. 1968: Monthly and seasonal mean global charts of brightness from ESSA 3 and ESSA 5 digitized pictures, Feburary 1967–Feburary 1968, ESSA Technical Report NESC 46.

Wada, H., Asakura, T. 1967: Some relations between the behavior of the polar vortex and long-range weather forecasting. *WMO Technical Note* **87** 292–303.

DISTRIBUTION AND VARIATION OF CLOUDINESS AND PRECIPITABLE WATER DURING THE RAINY SEASON OVER MONSOON ASIA

Tadashi Asakura

Abstract: The distribution of cloudiness observed by the meteorological satellite ESSA was analyzed from the viewpoint of dynamic climatology during the Baiu season of 1969. The mean cloudiness over Monsoon Asia undergoes little change and is nearly 50%, but the monsoon cloud belt varies in time and space in accordance with the seasonal shift of the strongest westerly flow axis at the 500 mb level. The summer monsoon clouds form a belt in which a major amount of water vapor is transported, extending from the Bering Sea to South China. Monsoon rains occur in Japan when the belt of monsoon cloud is over Japan and ends when the cloud belt passes beyond. The total amount of precipitable water in East Asia increases with the variation of upper westerly flow. An increment of precipitable water is in general associated with a low index circulation pattern, reflecting an invasion by the wet southerly flow. The center of maximum precipitable water is found over South China in May, where a wet tongue extends eastward off the south coast of Japan. The wet center in China shifts northward from May to July in accordance with the northward advance of the monsoon rain zone.

1. INTRODUCTION

The cloud distributions during the summer monsoon period have been dealt with intensively by meteorologists and climatologists in East Asia. The first systematic observation of clouds was made by Agematsu (1939) for aeronautical meteorological purposes. Subsequently, Ishihara *et al.* (1947), Matsumoto *et al.* (1966, 1967, 1968) and Tsuchiya (1969) studied the morphology of the winter monsoon clouds around Japan and pointed out the importance of mesoscale activities in the formation of clouds and heavy snowfall.

The first analysis of winter monsoon clouds based on TIROS pictures was made by Fujita (1965), this was followed by a more detailed analysis by Tsuchiya (1966) on the orographical effects on the distribution of winter monsoon clouds over and around Japan. Tsuchiya and Fujita (1967) proposed the criterion of mean vertical wind shear for three different cloud patterns in a detailed study of winter monsoon clouds over the western Pacific.

Studies of the summer monsoon clouds are much scarcer and only the case of heavy precipitation in the Baiu season has been studied by Tsuchiya and Hoshina (1966). Tuller (1968) analyzed the world distribution of precipitable water, using 182 stations throughout the world. Arakawa (1952) discussed the precipitable water in Japan.

In this paper, the patterns of the summer monsoon clouds and the precipitable water were analyzed for the Baiu season, the rainy season in Japan, and the Mai-yü season, the rainy season in China, in early summer.

2. MEAN STATE OF CLOUDINESS DURING THE SUMMER MONSOON OVER ASIA

A daily record of cloudiness is available from photographs taken by the meteorological satellite ESSA. In this paper, the space-means of cloudiness are evaluated for every square of 2.5° longitude and 2.5° latitude over East Asia and the Northwest Pacific ranging from 100°E to 180°E and from 6°N to 60°N. The cloudiness in each square is assessed as 0 (cloud amount below 33%), 1 (34–68%) and 2 (above 67%), and these values are summarized for five days as the basic data. This is then divided by 10 and expressed in percentages. The data used in this paper are for the period from May 1 to July 29, 1969.

The mean cloudiness during the summer monsoon over Asia is 47% and its time variations are only several percent, as shown in Fig. 1. Over East Asia (100–140°E), the mean cloudiness has a maximum value of 54% on June 25–29, when the Baiu rain was heaviest, and a minimum value of 40% on July 20–24 when the rainy season ended and the true summer season arrived. Over the Northwest Pacific Ocean (140–180°E), maximum (54%) and minimum (40%) values were observed on May 11–15 and 21–25. The average cloudiness over the region shows small variations in time, but large variations in space, as discussed in the next section.

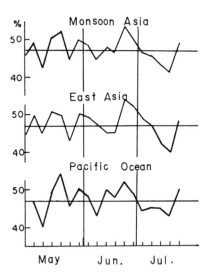

Fig. 1.   5-day mean cloudiness over East Asia (100–140°E) and the Northwest Pacific Ocean (140–180°E) and their means.

a.   Isopleth of 5-day mean cloudiness

The 5-day mean cloudiness averaged along every 2.5° latitude zone over East Asia from May to July is shown in Fig. 2a. This figure indicates that a heavy cloud zone is formed south of 30°N latitude in May and is displaced northward in June and July. South of this cloud zone, a low cloud zone is formed and is also displaced northward. The northward displacement is closely followed by the seasonal shift of the strongest axis of the westerly flow at the 500 mb level. The summer monsoon rain (Baiu) in Japan becomes active when the monsoon cloud belt reaches 30°N, and the summer season with hot, moist, fine weather arrives when the clear zone reaches 30°N latitude.

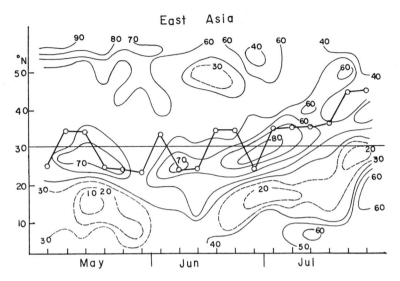

Fig. 2a. 5-day mean cloudiness for every 2.5° latitude over East Asia. The full line means the axis of the storongest westerly flow at the 500 mb level.

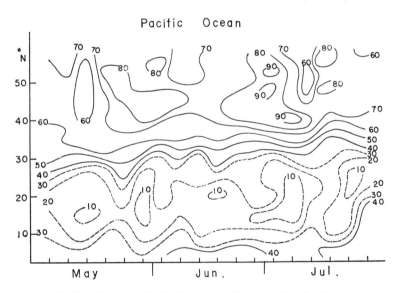

Fig. 2b. Same as Fig. 2a but for the Northwest Pacific Ocean.

On the other hand, the 5-day mean cloudiness over the Northwest Pacific Ocean does not change in time, as shown in Fig. 2b. The low cloud zone remains stationary south of 30°N while the heavy cloud zone is also stationary north of 30°N. The former corresponds to the Pacific anticyclone and the latter to the Aleutian low. Accordingly, the northward displacement of the summer monsoon clouds seems to be affected by the atmospheric circulation over East Asia.

b. *Distribution of 15-day Mean Cloudiness*

To make clear the large-scale features of the cloud distribution, four periods of 15-day mean cloudiness were analyzed. Each chart shows the pronounced cloud belt running from the Aleutian Islands to South China and a zone with low cloudiness over the Pacific Ocean. The cloud belt and the zone with low cloudiness shift northward with time as shown in Figs. 3a–d.

The characteristics of cloud distribution in each of the four charts are summarized as follows:

1) Before the rainy season (May 31–June 14, 1969)

The cloud distribution before the rainy season is shown in Fig. 3a. In this figure, the two cloud bands are seen south and north of Japan, forming a double structure of cloud band in East Asia. The northern cloud band running from the Sea of Okhotsk to Manchuria corresponds to the Eurasian polar frontal zone, and the southern one to the Pacific polar frontal zone pointed out by Yoshino (1969).

The southern cloud band, the cloud band with the summer monsoon, originates over Yunnan in South China and extends eastward off the southern coast of Japan. Eventually it combines with the cloud bands of the Aleutian low. North of the summer monsoon clouds a zone of low cloudiness is found, corresponding to the zone of moving anticyclone. South of the summer monsoon cloud zone, there is an expanding zone of clear sky which corresponds to the zone of North Pacific anticyclone. The monsoon clouds cover Yunnan, Taiwan and the Ryukyus in this period. This means that the summer monsoon rain, the rainy season in early summer called Mai-yü in China or Baiu in Japan, starts in these areas.

2) The first half of the rainy season (June 15–29, 1969)

Nearly the same features of cloud distribution as in Fig. 3a are found in Fig. 3b, which shows the start of the Baiu season in Japan. The summer monsoon clouds shift slightly northward over South China and considerably over the Pacific Ocean. On the other hand, the cloud zones along the Pacific polar frontal zone and the Aleutian low (the Eurasian polar frontal zone) do not change very much.

The summer monsoon forms a marked cloud zone covering Yunnan, the Yangtze, the Ryukyus and the Pacific Ocean side of Japan. This fact suggests that the Mai-yü in China and the Baiu in Japan are caused by a common cloud cover. When the monsoon cloud zone reaches the Pacific Ocean side of Japan, the Baiu season has started.

3) The latter half of the rainy season (June 30–July 14, 1969)

The cloud zone formed by the summer monsoon moves north over China and Japan, while it remains stationary over the Bering Sea. North China, the East China Sea, South Korea and Japan are under the influence of this zone. The Baiu is active when a thick cloud covers Japan, as seen in Fig. 3c.

The clear zone over the Pacific Ocean extends west to Vietnam. The centers of the clear sky are located northeast of the Phillipines and east of Marcus Island. South of this clear zone, there is a small cloud band over the southern Phillipines. This cloud band seems to corresponded to the I.T.C.

4) Summer season (July 15–19 1969)

The summer monsoon clouds disappear over East Asia when the summer season comes, as shown in Fig. 3d. But separate areas of high cloudiness are still seen over Korea, Manchuria, inland China and the Aleutian Islands. The first two zones are thought to be the remains of the summer monsoon clouds while the latter is part of the cyclone clouds with the Aleutian low.

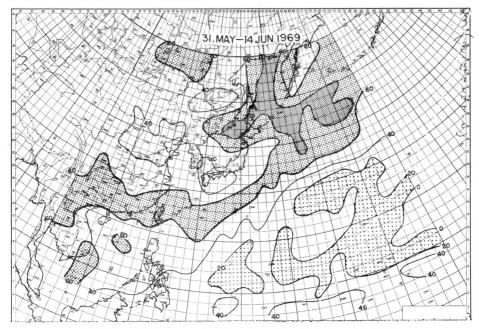

Fig. 3a. 15-day mean cloudiness from May 31 to June 14, 1969.

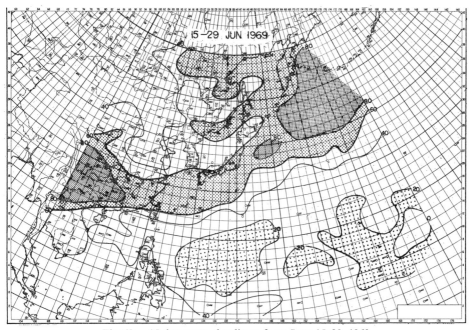

Fig. 3b. 15-day mean cloudiness from June 15–29, 1969.

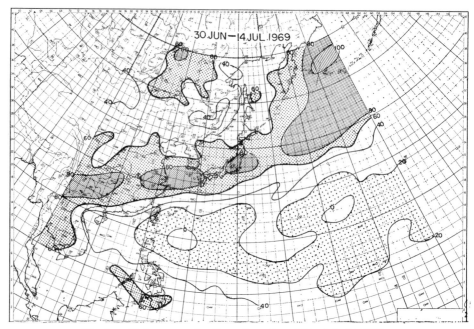

Fig. 3c.   15-day mean cloudiness from June 30 to July 14, 1969.

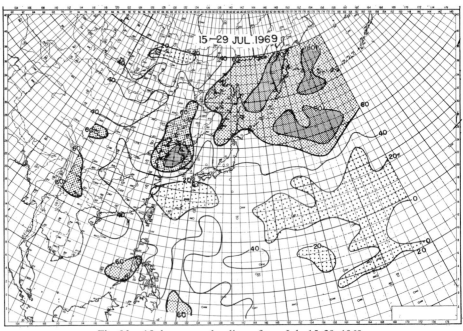

Fig. 3d.   15-day mean cloudiness from July 15–29, 1969.

South of Japan, a vast area of clear sky is found, corresponding to the anticyclone over the Pacific Ocean. This cloudless zone extends westward, covering China after the summer monsoon clouds retreat to the north. The northward advance of cloudless zone is remarkable, especially over Japan where the summer is hot, moist and clear.

Near the Phillipines and in Yunnan isolated cloudy areas are seen. In terms of the cloud distribution, the I.T.C. runs southwest of the Phillipines and extends northwest, but it is cut down over the South China sea. Similar conditions for the I.T.C. in July were found also in the figure presented by Yoshino (1969).

3. Northward Shift of the Summer Monsoon Cloud Zone, the Climate of Japan

As mentioned in Section 2, the summer monsoon cloud zone shifts northward during May, June and July. The climate of Japan is dealt with in connection with the behavior of the summer monsoon cloud zone.

A profile of 5-day mean cloudiness averaged along latitudes over East Asia is shown in Fig. 4, in which A shows the zone of summer monsoon clouds and B the clear sky zone of the subtropical anticyclone. As seen in Fig. 4, A and B advance to the higher latitudes during June and July. When A reaches close to 25°N, the Baiu season sets in. When A reaches 30°N, the rainfall is at its most severe for the Baiu season. B also shifts to the higher latitudes, from around 10 to 20°N, during the latter half of June. When A advances north of 30°N, the Baiu ends in Japan. When B reaches 30°N, fine, moist and hot summer weather prevails in Japan.

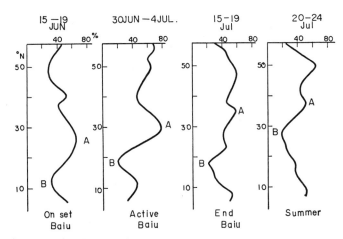

Fig. 4. 5-day mean cloudiness profile averaged over East Asia. The ordinate is the latitude and the abscissa cloudiness in percentages.

This behavior of the summer monsoon cloud zone is also reflected in the cloudiness distribution. For example, the average cloudiness along 30°N and the position of the low cloud zone along 130°E are shown in Fig. 5. Pre-Baiu conditions appear in Japan in the middle of May when the average cloudiness at 30°N increases to more than 60% and the position of the low cloud zone is north of 20°N. In accordance with the southward shift of the low cloud zone, the average cloudiness at 30°N decreases during the first half of June. Again, the low cloud zone moves northwards from the middle of June to the end of July.

When the low cloud zone is situated between 20–30°N, the average cloudiness at 30°N increases again to more than 60%, corresponding to the active period of the Baiu season. But in July the low cloud zone advances north of 30°N and the average cloudiness at 30°N decreases rapidly, resulting in the beginning of the summer season in Japan.

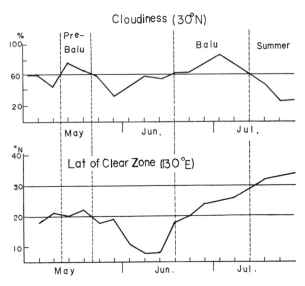

Fig. 5.   5-day mean cloudiness at 30°N averaged over East Asia (upper) and the latitude of the clear zone along 130°E (lower).

4.   Relation between the Cloud Distribution and the Synoptic Pattern

The distribution of rainfall in East Asia has a close connection with the upper air flow pattern (Asakura, 1968). The summer monsoon cloud zone also has a close relationship with the synoptic pattern. In this section, the morphological relations between the two are examined.

a.   *Relation between the Summer Monsoon Cloud and the Transport of Water Vapor*

The vector sum of water vapor transportation at the 1,000, 850, 700, and 500 mb levels was calculated with the electronic high speed computer HITAC 5020F. The length of the arrow in Figs. 6a and 6b is proportional to the amount of transport of water vapor.

The 5-day mean cloudiness and the water vapor transport are shown in Fig. 6a (May 26–30) and Fig. 6b (July 7–12). In Fig. 6a, the two strong water vapor flows are found in the outer region of the Pacific anticyclone and over the South China Sea crossing Vietnam. Both water vapor flows converge over the South China Sea, Taiwan, and the Ryukyus, where the summer monsoon cloud zone lies. In other words, the summer monsoon cloud zone is formed where the water vapor transportation is strong and the wet flows coming from the Pacific Ocean, Vietnam and China converge. During the active period of the Baiu season, the wet flow is very strong, originated mainly from the Southwest Pacific Ocean as shown in Fig. 6b. This finding has already been pointed out by Murakami (1959) and

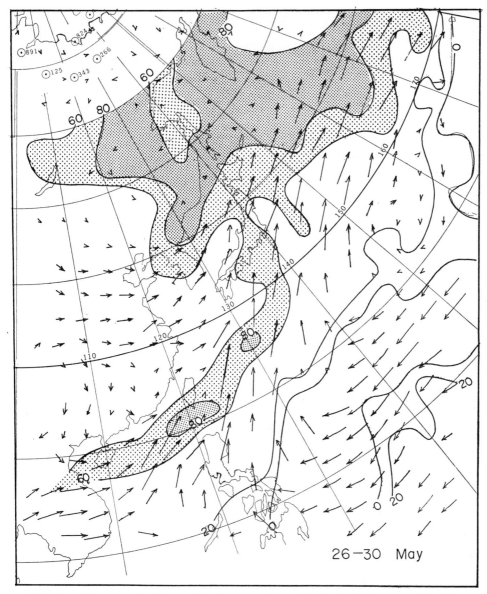

Fig. 6a. 5-day mean cloudiness and water vapor transport before the Baiu season (May 26–30, 1969).

Saito (1966). A very strong wet flow, streaming over the East China Sea and the southern part of Japan, is observed where the precipitation is heavy.

b. *Relation between the Cloud Distribution and the Synoptic Pattern*

The 5-day mean cloudiness is superimposed on the 5-day mean 500 mb and the surface pressure chart (Fig. 7). As shown in Fig. 7, the cloudiness distribution is more closely related to the 500 mb pattern than to the surface pressure pattern. For example, the summer monsoon clouds are formed even in the region of surface anticyclone over Central China

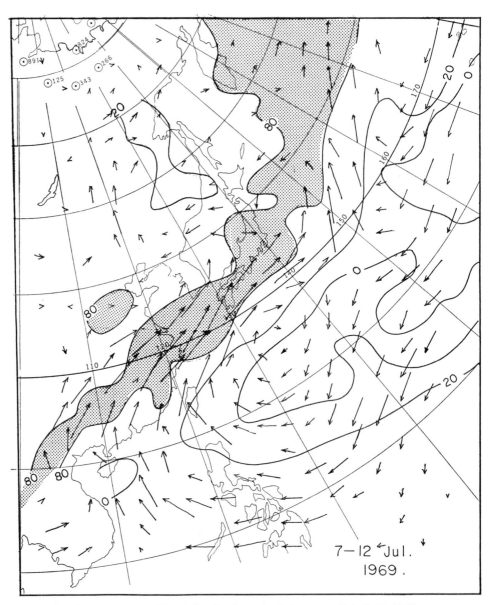

Fig. 6b. Same as Fig. 6a but for the active Baiu season (July 7–12, 1969).

and the Pacific Ocean east of Japan. At the 500 mb level, the summer monsoon clouds are formed along the strongest westerly axis where the southwesterly flows prevail. The clear zone is found over the Pacific anticyclone at the 500 mb level.

In a similar way, the 5-day mean cloud distributions are superimposed on the 5-day mean 500 mb charts in the pre-Baiu period, the onset of Baiu, the active Baiu and the end of Baiu as shown in Fig. 8. As in the case of Fig. 7, the zone of summer monsoon clouds is formed along the axis of the southwesterly and the low cloud zone is seen in the region of the Pacific anticyclone. In the pre-Baiu season (May 11–15), the cloud zones are formed south

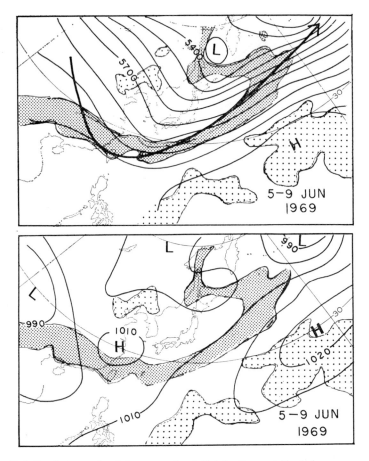

Fig. 7. Relation between the 5-day mean cloud distribution and the 5-day mean synoptic pattern at 500 mb (upper) and surface pressure (lower).

of the westerly axis, but at the start of the season, the cloud zones are seen just under the westerly, where the southwesterly flow is converging. In the active Baiu season (June 30–July 4), the axis of the southwesterly runs over Japan, and the monsoon clouds are also under the axis. But the monsoon cloud over China is separate from the westerly axis, which is directed northwestward; that is, the summer monsoon cloud band is formed only along the wet southwesterly flow. At the end of the Baiu season, the westerly axis shifts north of Japan and areas of clear sky are found over Japan and the Central Pacific, where the anticyclone develops. The cloud zones distribute themselves separately along the westerly axis.

5. PRECIPITABLE WATER OVER ASIA IN THE BAIU SEASON

The precipitable water at 44 stations over East Asia was calculated from upper-air data. The distributions of the stations and the area used are shown in Fig. 9. The isoline of the precipitable water is drawn from the daily values in May, June and July of 1968. The amount of precipitable water at the grid points of every intersection at 5° latitude and 5° longitude

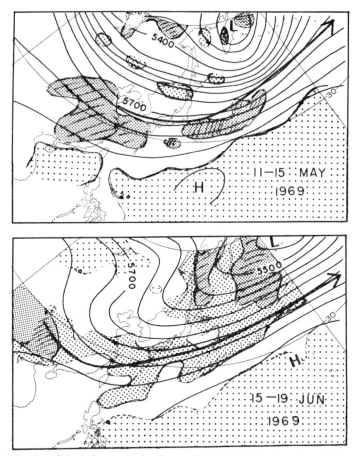

Fig. 8. Behavior of the 5-day mean cloudiness chart superimposed on the 5-day mean 500 mb pattern during the Baiu season. Pre-Baiu (May 11–15), onset of Baiu (June 15–19), active Baiu (June 30–July 4) and end of Baiu (July 15–19).

was read out by interpolation and the total precipitable water over East Asia was then estimated from the read out data.

The precipitable water in East Asia was found to roughly double from May to July as shown in Fig. 10. More precisely, the precipitable water fluctuates around the mean values

Table I. Means (M) and standard deviations of the precipitable water in May, June and July 1968 (unit: 10^6mb · g/kg · km^2).

Area		Asia	Siberia	China	Japan	Okhotsk
May	M	1,328.8	318.5	358.7	36.4	14.4
		127.0	75.2	24.4	7.6	3.0
June	M	1,761.7	519.7	292.0	45.4	22.0
		272.1	111.8	40.5	7.8	3.7
July	M	2,438.7	801.2	351.9	63.0	26.2
		167.2	76.9	27.6	10.1	4.5

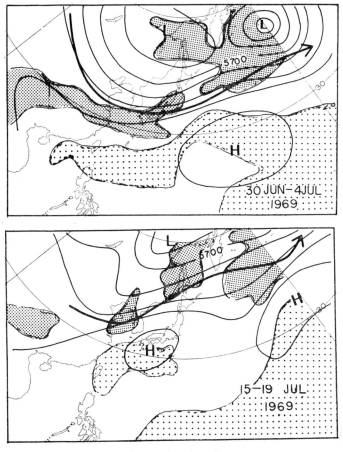

Fig. 8. (continued)

and its standard deviations are ± 127.0 (May), ± 272.1 (June) and ± 167.2 (July), as shown in Table I.

Each area has its maximum value in July, but the standard deviations are largest in June over Siberia and China, and in July over Japan and the Sea of Okhotsk. It is interesting to note that the time variation of the precipitable water has a close relationship to the frontal activity during the Baiu season. For example, the precipitable water increases rapidly during the pre-Baiu period (May 20) and then decreases at the end of May. It increases remarkably at the beginning of the Baiu season, and again towards the end of June and the first half of July at the climax of the Baiu season. It attains a maximum value in about the middle of July when the Baiu season is over in Japan, and then decreases gradually after the true summer season begins.

The behavior of the precipitable water over China, Japan, Siberia and the Sea of Okhotsk is shown in Fig. 11. In China, the variation is similar to the whole of Asia. But compared with the variation of zonal index (40–60°) at the 500 mb level in Asia, a reverse relation is found. This means that the strong westerly flow brings dry air into the continent, while in the period of weak westerly flow the meriodional flow prevails, resulting in an increment of

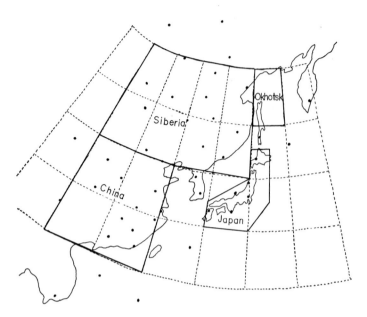

Fig. 9. Distribution of upper air observations and area of calculation of the precipitable water.

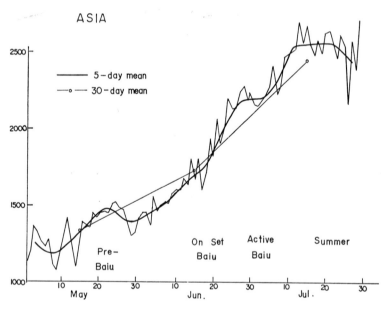

Fig. 10. Precipitable water summarized over Asia in May, June and July. Unit: 10^6 mb · g/kg · km².

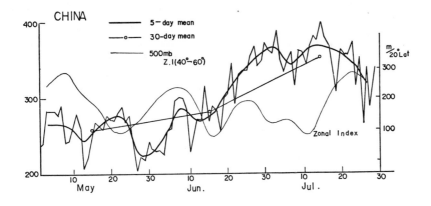

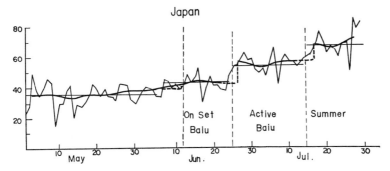

(a)   China (upper) and Japan (lower).

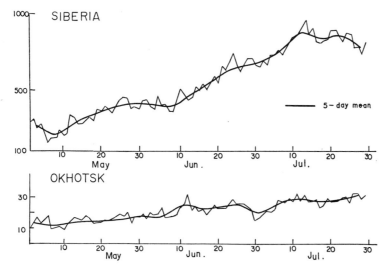

(b)   Siberia (upper) and the Sea of Okhotsk (lower).

Fig. 11. Precipitable water summarized over China, Japan, Siberia and the Sea of Okhotsk in May, June and July. Unit: 10^6mb $\cdot$ g/kg $\cdot$ km^2.

precipitable water. In other words, the precipitable water increases when the circulation is in a low index.

In Japan, the time change of the precipitable water does not resemble that of Asia or China as shown in Fig. 11(a). It increases discontinuously from one level to another. It is interesting to find that each level has some meaning. The first level corresponds to the onset of Baiu, the second to the most active period of Baiu and the third to the summer season. This fact will be utilized for the classification of seasons in Japan.

The daily and 5-day mean precipitable water in Siberia and the Sea of Okhotsk are shown in Fig. 11b. In Siberia, the mode of variation is similar to that in Asia as a whole. In the Sea of Okhotsk, on the other hand, the precipitable water increases very slightly and remains nearly constant. This suggests that the summer monsoon does not invade the Sea of Okhotsk.

6. THE DISTRIBUTION OF PRECIPITABLE WATER

In this section, the distribution of precipitable water during the Baiu season is discussed and the relation between the wet tongue and rainfall activity is analyzed.

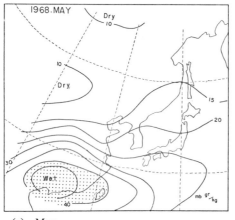

(a) May.

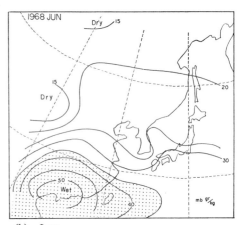

(b) June.

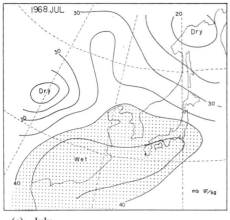

(c) July.

Fig. 12. Distribution of monthly mean precipitable water in May, June and July, 1968.

a. *Distribution of Monthly Mean Precipitable Water in May, June, and July*

The distribution of monthly mean precipitable water over Asia in May, June, and July is shown in Fig. 12. In May (Fig. 12a), the wet center is in South China and has a value of 45 mb g/kg. The wet area extends east off the southern coast of Japan. Over Siberia, the precipitable water is only one-fourth of the wet center.

In June (Fig. 12b), the wet center is located in the same position as in May, though the value increases from 45 to 50 mb g/kg. Water vapor is transported northward over Siberia where the precipitable water increases by about 10 mb g/kg. The isoline of 20 mb g/kg moves northward from about 40° to 50°N. The wet zone over South China extends eastward, increasing in intensity, and the isoline of 30 mb g/kg proceeds over Japan, though it remains stationary in China during May and June. The Baiu season in Japan is caused by this invasion of wet air.

In July (Fig. 12c), the wet center proceeds northward but its value decreases slightly. The wet zone above 40 mb g/kg covers only South China (May) and the Ryukyus (June), but in July the wet zone expands over a very large area covering South and Central China, Korea, and Japan. The precipitable water over the Pacific Ocean side of Japan has nearly the same value as the wet center over China, suggesting that both the Baiu and the Mai-yü are part of the summer monsoon rain complex in Asia.

Over Siberia, a remarkable increase of precipitable water is found. The isoline of 30 mb g/kg runs over Japan and Central China in June, but runs over Siberia in July. That is, Siberia in July is nearly as wet as Japan in the first half of the Baiu season. On the other hand, dry weather persists over the Sea of Okhotsk as mentioned in the previous section.

b. *The Variation in Precipitable Water and the Change in Baiu Activity*

Since the Baiu season is the wettest season in Japan, the weather during this period is directly affected by the variation in precipitable water. Four examples are shown in Figs. 13a–d.

1) Onset of the Baiu season

Fig. 13a shows the distribution of precipitable water on the first day of the Baiu season. Very wet air lies South China and North Vietnam, and a wet tongue extends to Japan, passing through Taiwan and the Ryukyus. The axis of the wet tongue coincides well with the position of the surface front south of Japan. The wet air covers not only Japan, but also Siberia. The latter wet area is brought about by a southerly flow accompanied by an upper-air depression.

On the other hand, dry air comes down to the east of Japan from the cold Sea of Okhotsk and forms a front between the wet tongues. Another area of dry air originating in North China occupies East China Sea and forms a very distinct contrast to the wet tongue over the Ryukyus.

The onset of the Baiu season is characterized both by the invasion of very wet air into Japan and the strong intrusion of dry air over East China Sea and over the Pacific Ocean east of Japan.

2) Break in the Baiu season

Fig. 13b shows the distribution of precipitable water on a day when the Baiu season breaks. The wet center is over South China, where one wet tongue extends to the north, covering North China and Siberia, and the other extends along the front south of Japan.

Very dry air invades the Japan Sea and the East China Sea and their values are smaller

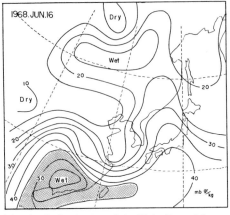

(a) At the beginning of the Baiu (June 16, 1968).

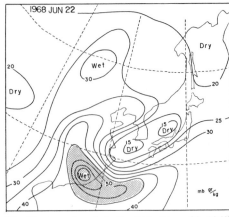

(b) During a break in the Baiu (June 22, 1968).

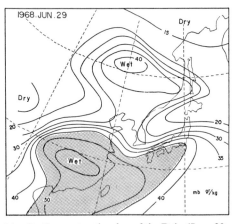

(c) On the most active day of the Baiu (June 29, 1968).

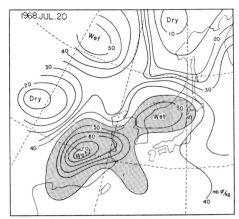

(d) At the end of the Baiu (July 20, 1968).

Fig. 13. Distribution of precipitable water in the Baiu season (1968).

than the one over the Sea of Okhotsk. This intrusion of dry air brings a break in the Baiu season.

3) Active Baiu season

Fig. 13c shows the distribution of precipitable water on a day of very heavy precipitation in Japan. The wet center is along the Yangtze and the wet tongue extends to Japan. The amount of precipitable water on the Pacific Ocean side of Japan is the same as that over China. This causes the heavy precipitation. Another wet center lies over Siberia, with a value of 40 mb g/kg, which is nearly the same as over the monsoon area.

On the other hand, dry air extends over North China and the Japan Sea. This forms a belt of dry air between two wet zones, and creates sharp contrast in wetness along Japan, the East China Sea and Central China. The heavy precipitation was observed over the zone with this sharp contrast of wetness.

4) End of the Baiu season

Fig. 13d shows the distribution of precipitable water at the end of the Baiu season. The wet air covers Central China, Korea, and the Japan Sea. Relatively dry air lies over the Pacific Ocean and Japan. This means that the monsoon rains end over Japan and start over Korea. The dry zone moves northward and spreads inland over North China, Manchuria and the Sea of Okhotsk. Further north a wet center with 50 mb g/kg is observed. This center as well as the center shown in Fig. 13c is important in considering the the water balance in southeastern Siberia and Mongolia.

To observe more clearly the behavior of the precipitable water at the begining and the end of the Baiu season, the specified isolines of 35 and 45 mb g/kg are traced back as shown in Fig. 14. On June 6, the isoline of 35 mb g/kg is over South China and progresses northeast, that is, it is near Taiwan on the 8th, the Ryukyus on the 10th, the southern coast of Japan on the 13th and over Japan on the 16th when the Baiu starts in Japan. Accordingly, the Baiu in Japan seems to be brought about by an invasion of the wet monsoon current from South China or its vicinity.

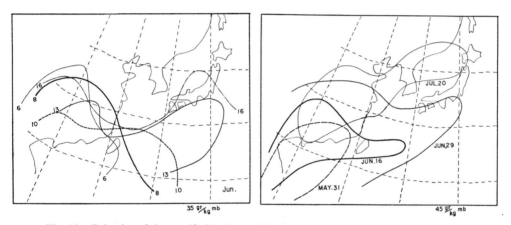

Fig. 14. Behavior of the specified isolines of 35mb · g/kg (left) and 45mb · g/kg (right).

In a similar fashion, the isoline of 45 mb g/kg moves northeast. On the 16th the isoline moves over the Ryukyus, on the 29th the isoline advances over the southern coast of Japan, when heavy precipitation occurs, and on July 20th the isoline moves to the Japan Sea when the Baiu season ends in Japan. It can therefore be said that the behavior of the moisture is a good indicator of the mode of the summer monsoon.

7. CONCLUSION

Though this is a case study, the results we obtained can be interpreted in a more generalized fashion as follows:

1) The Mean cloudiness over Asia is 47% during the summer monsoon and its time variation ranges only over several percent. But the cloud distribution shows a large variation in time and space over East Asia.

2) There is a stationary zone with heavy clouds over the Aleutian Islands and a non-stationary cloud belt over East Asia.

3) The latter is the cloud zone formed by the summer monsoon and shifts from south to north in accordance with the advance of the rainy season.

4) The Baiu season starts when the summer monsoon cloud belt approaches Japan. The Baiu front is active when the cloud belt runs over Japan, and ends when the cloud belt shifts to the Japan Sea.

5) In summer monsoon cloud belt, there is much transport of water vapor. The cloud belt shifts with the axis of the strongest westerly flow at the 500 mb level.

6) South of the summer monsoon cloud belt, there is a vast area of clear sky which also shifts northward in association with the summer monsoon season.

7) Another cloud belt is formed south of the clear zone, south of the Phillipine Islands, corresponding the I.T.C.

8) The center of precipitable water is located over South China in May and June and shifts over Central China in July.

9) The zone of precipitable water extends northeastward and invades Japan in the Baiu season.

10) The variation in precipitable water corresponds to the rainly and occasionally muggy weather of the Baiu and the hot, moist and fine weather of the true summer season over East Asia.

ACKNOWLEDGEMENTS

The author wishes to thank Prof. M. M. Yoshino, Dr. H. Wada and other members of the research group on the water balance of Monsoon Asia for their valuable discussions and suggestions. The author is also indelated to Mr. M. Matsushita and Miss R. Shitihyo for their help in preparing this manuscript.

References

Agematsu, K. 1939: Aeronautical study on the Japan Sea. *Report of Aeronautical Meteorology* **3** 202–250.
Arakawa, H. 1952: The precipitable water in Japan. *J. Met. Soc. Japan* **30** 203–209.
Asakura, T. 1968: Dynamic climatology of atmospheric circulation over East Asia centered in Japan. *Met. Geoph.* **19**(1) 1–68.
Ishihara, K., Toyama, Y., Okada, K., Koike, S. 1947: Report of the meteorological and oceanographical observations in the middle Japan Sea between Niigata and Nashin, North Korea. *Hokuriku Kenkyu Kaishi* **2** 213–225.
Fujita, T. 1965: Influence of the Japanese Islands upon the winter monsoon clouds. *JMA Tech. Rept.* **47** 247–267.
Matsumoto, S. Ninomiya, K. 1966: Some aspects of cloud formation to heat and moisture supply from the Japan Sea under the weak winter monsoon. *J. Met. Soc. Japan* **44** 60–75.
Matsumoto, S., Ninomiya, K., Akiyama, T. 1967: Cumulus activities in relation to the mesoscale convergence field. *J. Met. Soc. Japan* **45** 292–305.
Matsumoto, S., Ninomiya. K., Akiyama, T. 1968: Mesoscale analytic study on a lined up cumulus row caused by orographic effect under winter monsoon situation. *J. Met. Soc. Japan* **46** 222–233.
Murakami, T. 1959: The general circulation and water-vapor balance over the Far East during rainy season. *Geophys. Mag.* **29** 131–171.
Saito, N. 1966: A preliminary study of the summer monsoon of southern and eastern Asia. *J. Met. Soc. Japan* **44** 44–59.
Tsuchiya, K. 1966: Weather Satellite—Weather as seen from space. *Gakugei Shobo Tokyo* 198.
Tsuchiya, K. 1969: The morphology of winter monsoon clouds around Japan. *Geophys. Mag.* **34** 427–445.
Tsuchiya, K., Fujita, T. 1967: A satellite meteorological study of evaporation and cloud formation over the western Pacific under the influence of the winter monsoon. *J. Met. Soc. Japan* **45** 235–250.

Tuller, S. E. 1968: World distribution of mean monthly and annual precipitable water. *Mon. Wea. Rev.* **96** 785–797.

Winston, J. S., Taylor, V. R. 1968: Monthly and seasonal mean global charts of brightness from ESSA 3 and ESSA 5 digitized pictures, February 1967–February 1968. ESSA Technical Robert NESC46.

Yoshino, M. M. 1969: Climatological studies on the polar frontal zones and the intertropical convergence zones over South, Southeast, and East Asia. *Climatol. Notes, Hosei Univ.* (1) 1–71.

SYNOPTIC AND CLIMATOLOGICAL STUDY ON THE UPPER MOIST TONGUE EXTENDING FROM SOUTHEAST ASIA TO EAST ASIA

Atsushi KURASHIMA and Yoji HIRANUMA

Abstract: Baiu, a rainy season in early summer over East Asia, is discussed from the standpoint of synoptic and aerological climatology. It can be said that the Baiu front is the polar front in Japan but the tropical front in China. This is also confirmed by the facts that the frontal zone during Baiu in the neighbourhood of Japan and the sea east of Japan is characterised by a sharp meridional temperature gradient and on the contrary, the frontal zone over China and the East China Sea is characterised by a sharp meridional gradient of water vapor. Then the tropical and subtropical monsoon zones are defined and their model at the 700 mb level is drawn. Finally the so-called moist tongue, circulation systems and anticyclones in the upper layer are dealt with.

1. BAIU AND BAIU FRONT

One of the characteristic features of the climate in Japan is the rainy season called the Baiu which occurs from early June to mid-July. The formation of this rainy season has been explained as stagnancy of the Baiu front in the neighbourhood of Japan and the activities of the extratropical cyclones on the frontal zone.

The area where the Baiu front is likely to stagnate, namely the climatological Baiu frontal zone, shifts seasonally (early summer—hot summer) from south to north in the vicinity of Japan (Yoshino, 1963, 1965). Corresponding to this movement, the rainy season also shifts northwards, and the prime of the rainy season appears generally in May in Okinawa and the Ogasawara Islands, in mid- and late-June in the southern part of Honshu, and in early and mid-July in the northern part of Japan.

It is recognized by synoptic meteorologists in Japan that there are roughly two types of rainfall during the Baiu season. One is the ordinary rain which falls also during the winter, and can be explained as the activity of extratropical cylones on the polar front. The other is the considerable amount of rain which falls for a comparatively short time and in a comparatively restricted area. Frequently the rainfall is as high as 200–300 mm/day, and locally it may reach 1,000 mm/day. This kind of rainfall is more closely related to a mesoscale disturbance than to extratropical cyclones or a synoptic scale front. Such a great amount of precipitation is generally observed in the tropical humid climate zone and tropical monsoon zone. Therefore, the heavy rainfall observed during the Baiu season suggests that the tropical air flows participate in the mechanism of the Baiu front, or at least a part of the Baiu front has a similar mechanism to the tropical convergence zone.

2. THE BAIU AND MAI-YÜ FRONTS FROM THE VIEWPOINT OF AIRMASS THEORY

The Russian meteorologist Woeikov was the first to suggest that there might be a relationship between the formation of the rainy season (Baiu in Japanese and Mai-yü in Chinese), appearing characteristically in the region extending from the basin of the Yangtze River toward Japan during the period from early June to mid-July, and the southwest or tropical monsoon. This idea was confirmed by Chinese meteorologists in the 1930's and 1940's through modern meteorological analyses (Collected Scientific Papers, Academia Sinica 1954). They concluded, using surface observational data, that the air flow which runs over the earth's surface on the south side of the Baiu front in the basin of the Yangtze River is the southwest monsoon (equatorial air mass), and therefore that the Mai-yü front in China is the tropical front. In Japan and the USSR, on the other hand, it has been thought that the Baiu front is the polar front between the maritime polar air mass in the Sea of Okhotsk and the northwest part of the northern Pacific, and the maritime subtropical air mass of the Pacific subtropical anticyclone (Stremousov, 1935).

The chinese meteorologists in the 1930's and 1940's reported that the Mai-yü in China in which the equatorial air mass (tropical monsoon) plays an important role is far more remarkable, in quantity and strength, than the Baiu in eastern Japan in which the tropical air mass (subtropical monsoon) plays an important role (Yao, 1939). Undoubtedly, the drizzle and moderate rain during the first half of the Baiu season in eastern Japan can be explained as the polar front. However, the severe rains of 200–1,000 mm/day, occurring mostly in western Japan during the latter half of the Baiu season suggest that the equatorial airmass is the chief moisture carrier. Figure 1 summarizes the opinions of various climatologists concerning the mechanism of the Baiu front, at the time when the Baiu was discussed from the standpoint of airmass theory (Kurashima, 1958b).

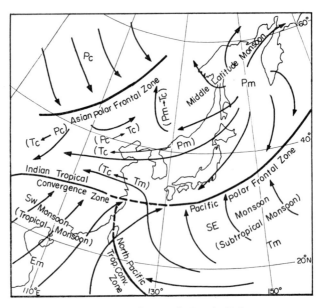

Fig. 1.   Main frontal zones during the Baiu season.

3. Upper Moist Tongue from the Synoptic Viewpoint

With the development of an upper air observation network in East and South Asia, the knowledge of the vertical sturcture of synoptic process in these areas increased rapidly. Thompson (1951) discussed the seasonal characteristics of the upper air system at the 10,000 ft level in Southeast Asia, and depicted the typical upper air system in each season as shown

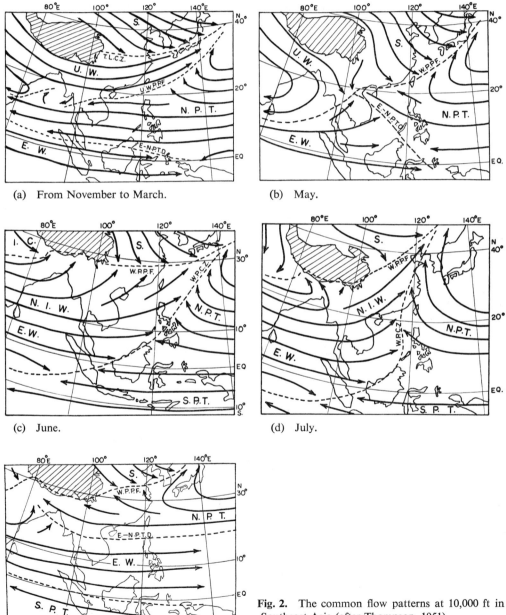

(a) From November to March.

(b) May.

(c) June.

(d) July.

(e) September.

Fig. 2. The common flow patterns at 10,000 ft in Southeast Asia (after Thompson, 1951).

in Fig. 2. The point to be noted is that he considered that the Indian westerlies or the SW monsoon flow into Central China and Southern Japan during the Baiu season. A similar idea is presented in the mean upper air flow chart prepared by Chinese climatologists on the basis of abundant upper air observational data (Staff Members, Academia Sinica 1957). Meteorologists and forecasters in Japan have ascertained that the air flow system mentioned above is the moist tongue extending from the southwestern region to the neighborhood of Japan (Otani, 1954).

Figure 3 shows an example of the seasonal northward advance of the upper moist tongue with the Baiu front.

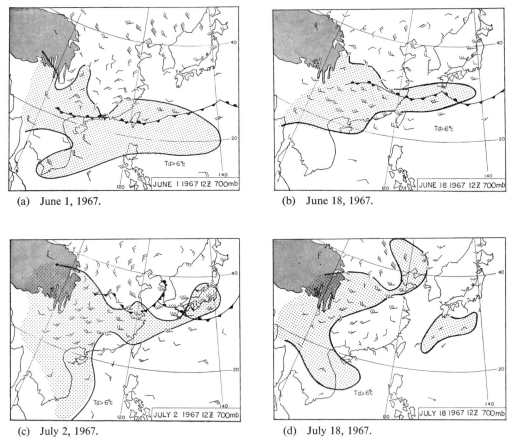

(a)   June 1, 1967.   (b)   June 18, 1967.

(c)   July 2, 1967.   (d)   July 18, 1967.

Fig. 3. Examples of the upper moist tongue at the 700 mb level. Shaded area: area with a value Td>6°C. Fronts refer to sea level.

4. Upper Moist Tongue from the Viewpoint of Climatology

The existence of the upper moist tongue and its seasonal shift, described in the previous section, is climatologically obvious. Figure 4 denotes the seasonal variation of the area where the monthly mean precipitable water is above 55 mm. This figure shows that in June, July, and August, moist air similar to that in Southeast Asia and India extends toward South

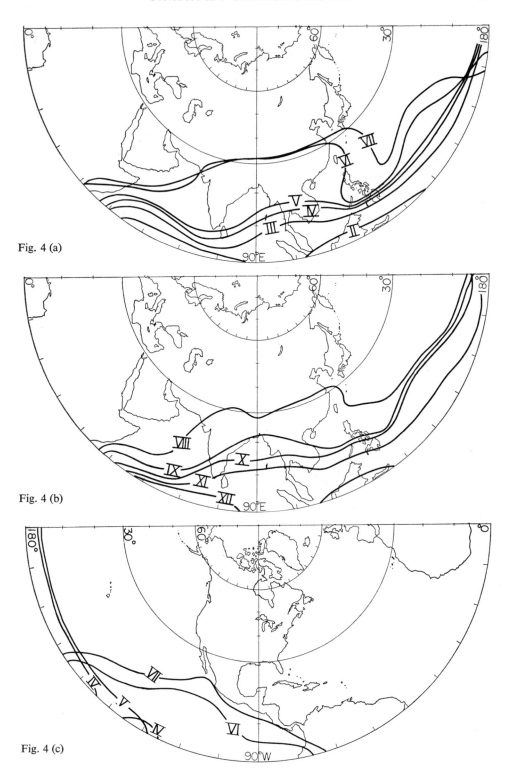

Fig. 4 (a)

Fig. 4 (b)

Fig. 4 (c)

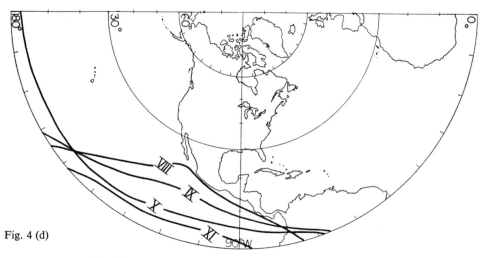

Fig. 4 (d)

Fig. 4a–d. Advance and retreat of the isopleth of monthly mean precipitable water of 55 mm. The Roman numerals indicate months.

China and the southwest seas of Japan. In North America, the seasonal northward expansion of moist air is not as remarkable as in East Asia. This fact suggests that the inflow of moist air from Southeast Asia and its adjacent seas into the middle latitudes of East Asia is one factor in the formation of the characteristic Asian monsoon.

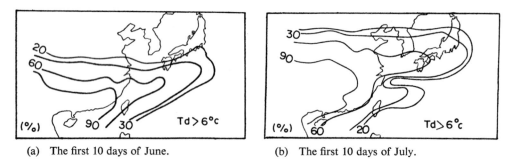

(a)   The first 10 days of June.   (b)   The first 10 days of July.

Fig. 5. Percentage frequency of dewpoint temperatures above 6°C at the 700 mb level (1960–1962). (After Akashi and Shitara, 1964.)

Figure 5 shows the frequencies of dewpoint temperature above 6°C at the 700 mb level in early June and early July. This figure indicates that the area in East Asia where the upper moist air is likely to appear makes a seasonal northward advance during the period from June to July (Akashi and Shitara, 1964).

5.   THE UPPER MOIST TONGUE SHOWN IN THE WATER VAPOR TRANSPORTATION CHART

As to the water vapor balance in East Asia, many studies have been made by Flohn (1950), Murakami (1959), Saito (1966) *et al.* Figures 6 and 7 are the mean water vapor transportation charts ($\overline{qv}$, where q is the specific humidity and v wind speed) for a period of 5

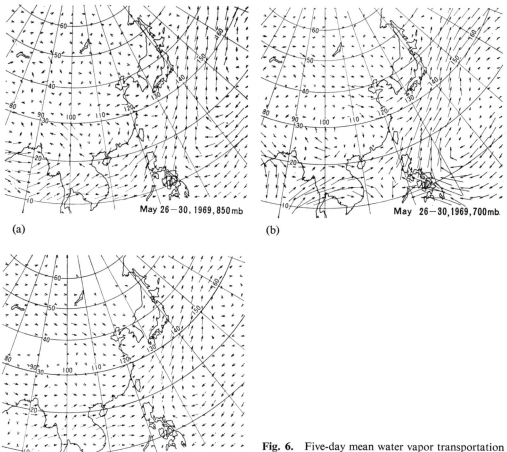

(a)

(b)

(c)

Fig. 6.   Five-day mean water vapor transportation just before the beginning of the Baiu season.

days in the rainy season in 1969. The quantitative study of the water vapor balance based on the data of these charts is presented by Asakura in this volume. Figures 8 and 9 denote the Baiu front at the surface and 850 mb levels for the corresponding period in Figs. 6 and 7, respectively. From these figures, we understand that in East Asia a considerable amount of vapor transportation comparable to that in the tropical region is observed along the Baiu front. Especially at the 500 mb level in Fig. 7, the mean amount of water vapor transportation from South China and the basin of the Yangtze River towards the southern coast of Japan is of the highest value. The vertical axis of the water vapor transportation maximum in East Asia is inclined to the north. This means that water vapor transportation in East Asia is related to the frontal activity.

6.   IS THE BAIU FRONTAL ZONE A POLAR FRONTAL ZONE OR A TROPICAL CONVERGENCE ZONE?

This problem was referred to in section 2, and it is sometimes discussed among Japanese meteorologists even now. For instance, Sugimoto (1967) showed that the frontal zone in the

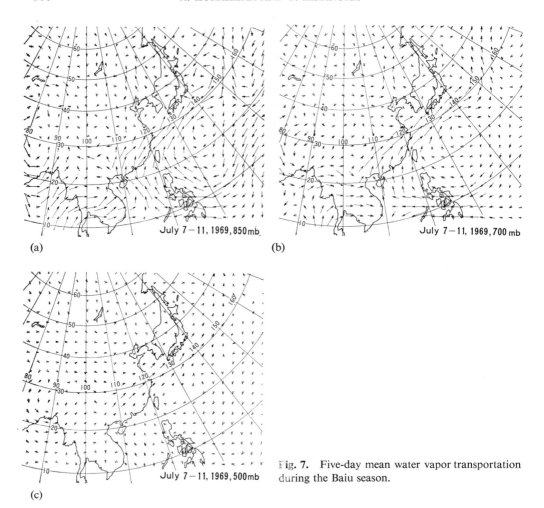

(a)   (b)

(c)

Fig. 7.   Five-day mean water vapor transportation during the Baiu season.

vicinity of Japan during the Baiu season has the nature of a subtropical frontal zone, not characterized by a high value temperature gradient, but by a high value gradient of the equivalent potential temperature containing a moisture factor. Figures 10 and 11 show the area with high values of meridional thickness (1,000–500 mb) gradient (full line) and the area with high values of meridional gradient of the mean precipitable water (dotted line) in January and June, respectively. These values have been calculated from the monthly means at each grid point of 10° lat. and 10° long.

The area with high values of meridional thickness gradient can be called the climatological polar frontal zone. The area with a high value of meridional gradient of precipitable water can be considered the northern limit of the inflow of moist air. In Fig. 11, it can be seen that Baiu frontal zone in the neighborhood of Japan and the sea to the east of Japan is characterized by a high meridional temperature gradient. In contrast, the Baiu frontal zone over China and the sea to the southwest of Japan is characterized by a high meridional vapor gradient. This fact tells us that the Baiu frontal zone in Japan and the sea to the east of Japan has relatively stronger characteristics of the polar frontal zone, while the Mai-yü front in China has relatively stronger characteristics of the tropical convergence zone.

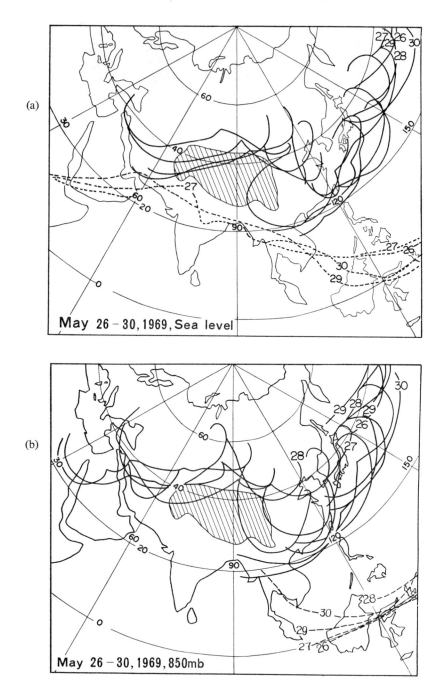

Fig. 8. Frontal system just before the beginning of the Baiu season (corresponding to Fig. 6). Full line: polar or actic front. Broken line: tropical front. Numerical figures indicate date.

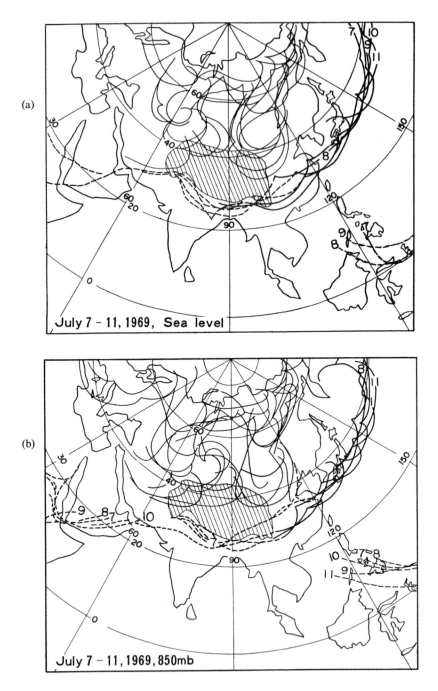

(a)

(b)

Fig. 9. Frontal system during the Baiu season (corresponding to Fig. 7). Full line: polar or actic front. Broken line: tropical front. Numerical figures indicate date.

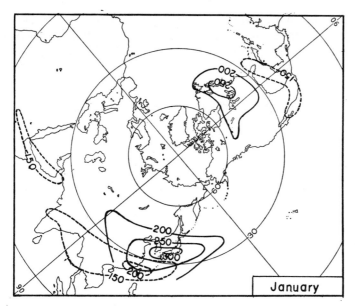

Fig. 10. Area with a high meridional gradient of thickness between 1,000–500 mb (full line) and area with a high meridional gradient of mean monthly precipitable water (dotted line) in January. Unit: m/10 lat. and 0.1 mm/10 lat. respectively.

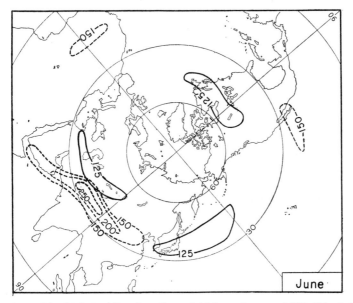

Fig. 11. Area with a high meridional gradient of thickness between 1,000–500 mb (full line) and area with a high meridional gradient of mean monthly precipitable water (dotted line) in June. Unit: m/10 lat. and 0.1 mm/10 lat. respectively.

However, depending on the year and on the season even in the same year, the Baiu frontal zone in the vicinity of Japan can have the characteristics of a tropical convergence zone like that in China and conversely the Mai-yü frontal zone in China can have the characteristics of a polar frontal zone like that in Japan.

7. MONSOON ZONES AND THE UPPER MOIST TONGUE

It was pointed out by Flohn (1960a) and Khromov (1950, 1956) that monsoons tend to develop along a latitudinal circle.

Kurashima (1959b) referring to Flohn and Khromov, demonstrated with the schematic models in Figs. 12 and 13 that the zonal features of the distribution of monsoon regions are due to the seasonal meridional movement of the zonal planetary wind belts. He named the region where the trade wind prevails in winter and the equatorial westerlies prevail in summer the "tropical monsoon zone; the region where the middle latitude westerlies prevail in winter and the trade wind prevails in summer the "sub-tropical monsoon zone;" and the region where the polar easterlies prevail in winter and the middle latitude westerlies in summer the "polar monsoon zone."

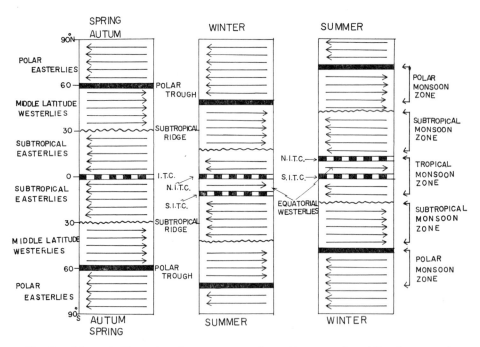

Fig. 12. Schematic illustration of monsoon zones formed by seasonal meridional movement of the planetary wind belts.

In East Asia, including Japan, the SE monsoon flowing from the Pacific subtropical anticyclone prevails in mid summer, namely in July and August. This wind is the turning current of the trade wind. It means that the monsoon in July and August in Japan is a subtropical monsoon.

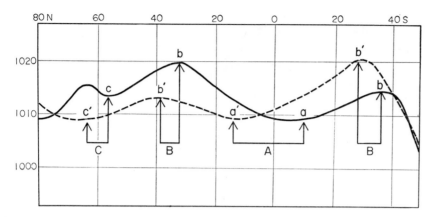

Fig. 13. Schematic illustration of monsoon zones formed by seasonal meridional movement of the planetary pressure system. Full line: Mean surface pressure of atmosphere plotted against latitude in January. Broken line: in August. a,a′: Equatorial trough, b,b′: Subtropical ridge. c,c′: Polar trough. A: Tropical monsoon zone. B: Subtropical monsoon zone. C: Polar monsoon zone.

The model of the monsoon zone at the 700 mb level (approx. 3 km high) prepared on the basis of much climatological data is shown in Fig. 14 (Kurashima, 1959).

This figure (Fig. 14) shows the following:

1) The planetary wind system at an approximately 3 km level consists of three systems: a. circumpolar westerlies or middle latitude westerlies, b. the wind system which forms the subtropical anticyclonic circulation, and c. equatorial westerlies.

2) These wind systems move northward from winter through summer. In this way, the tropical and subtropical monsoon zones are formed.

3) In summer, anticyclonic circulation is observed in the circumpolar westerlies. This upper anticyclone traverses eastward over Japan in late spring and early summer, and lingers

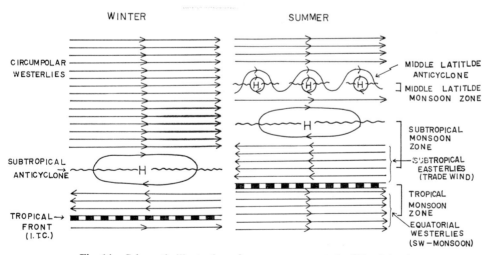

Fig. 14. Schematic illustration of monsoon zones at the 700 mb level.

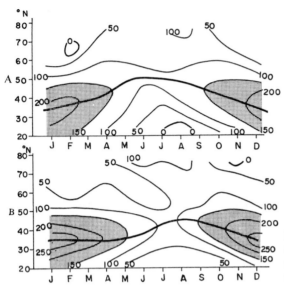

Fig. 15. Seasonal shift of climatological frontal zone. Isopleth of meridional thickness gradient along 100°E (A) and 150°E (B). Unit: m/10 lat.

in the vicinity of the Sea of Okhotsk during the Baiu season. And this anticyclonic circulation constitutes the summer middle latitude monsoon in contrast to the circumpolar westerlies in winter.

How can we explain the formation of the upper moist tongue in East Asia from the viewpoint of the monsoon zone model mentioned above? Kurashima (1968) explained that its formation is attributed to the difference in the seasonal northward displacement of the zonal planetary wind belt over the continent and over the ocean in early summer. Figure 15 shows the isopleth of meridional thickness gradient along 100°E (the Asian continent) and 150°E (the Pacific). The thick black lines indicate latitudes with maximum thickness gradient and show the seasonal shifting of the climatological frontal zone. Figure 16 is a comparison of seasonal shifting of the climatological frontal zone over the continent and the one over the Pacific. It shows the following:

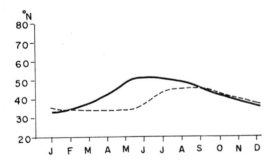

Fig. 16. Comparison of seasonal shift of the climatological frontal zone over the continent and the ocean. Full line: along 100°E. Broken line: along 150°E.

1) From March through June, the polar frontal zone over the continent moves northward, but the polar frontal zone over the Pacific is almost stationary.

2) The polar frontal zone over the Pacific moves northward from July through August. Fig. 17 shows that the intertropical convergence zone, that is, the monsoon trough, moves farther north over the continent than over the Pacific in early summer.

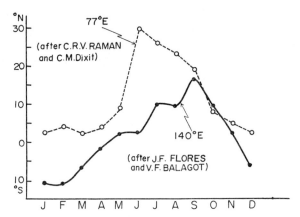

Fig. 17. Comparison of seasonal shift of the climatological intertropical convergence zone over the continent and the ocean.

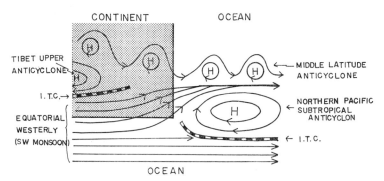

Fig. 18. Schematic illustration of 700 mb air flow pattern during the Baiu season.

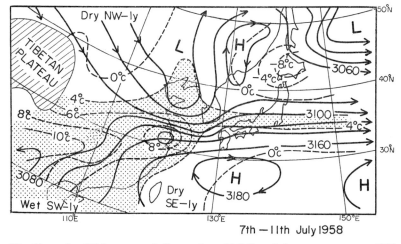

Fig. 19. Example of 5-day mean air flow pattern. Full line: 5-day mean contour of 700 mb level. Dotted line: isotherm of 5-day mean dewpoint temperature at 700 mb level. Shaded area: area with a value of Td>4°C.

These facts mean that in early summer the planetary wind system moves farther north over the continent than over the ocean. Hence, the summer upper air system on the right hand side in Fig. 14 also moves farther north over the continent than over the ocean. As a result, in a region where the western part is the continent and eastern part the ocean a wind system may be formed as shown in Fig. 18. The SW monsoon can flow into the area between the subtropical anticyclone and middle latitude anticyclone. During the Baiu season Japan is located between the subtropical and middle latitude anticyclones. On the Asiatic continent, especially in summer, the upper subtropical anticyclone is formed over the Tibetan Plateau by thermal factors (Flohn, 1957, 1960 b; Koteswaram, 1960; Murakami, 1958; Staff Members, Academia Sinica 1957, 1958). This anticyclone is located farther north than the Pacific subtropical anticyclone. This combination causes the formation of the wind system shown in Fig. 18.

The model in Fig. 18 resembles the upper air system during the Baiu season, which was obtained from actual weather charts by Thompson (1951) and Kurashima (1958 b, Fig. 19).

All these data tell us that, in order to understand the climate of Monsoon Asia, it is necessary to study the interaction of the tropical, subtropical and middle-latitude monsoons.

References

Akashi, A., Shitara, H. 1964: "Moist Tongue" over Japan in the Bai-u season (preliminary report). *Tôhoku Chiri.* **16**(4) 171–173 (in Japanese).

Flohn, H., Oeckel, H. 1956: Water vapour flux during the summer rains over Japan and Korea. *Geoph. Mag.* **27** 527–532.

Flohn, H. 1957: Large-scale aspects of the summer monsoon in South and East Asia. *75th Ann. Vol. J. Met. Soc. Japan* 180–186.

Flohn, H. 1960a: Monsoon winds and general circulation. *In* Monsoons of the World 63–74.

Flohn, H. 1960b: Recent investigation on the mechanism of the "summer monsoon" of southern and eastern Asia. *In* Monsoons of the World 75–88.

Flores, J. F., Balagot, V. F. 1969: Climate of the Philippines. *World Survey of Climatology.* **8** 159–213.

Khromov, S. P. 1950: Musson kak geograficheskaja realjnostj. *Izvestija Vsesojuznogo Geograficheskogo Obschchestva* **82** 225–246.

Khromov, S. P. 1956: Musson v obshchej tsuirkuljatsuii atmosfery. *In* Sovremennje problemj klimatologii. Gidrometeoizdat, Leningrad 84–106.

Koteswaram, P. 1960: The Asian summer monsoon and the general circulation over the tropics. *In* Monsoon of the World 105–110.

Kubota, I. 1969: Distribution of monthly mean precipitable water in the northern hemisphere and its time change. *Climatological Notes, Hosei Univ.* (2) 1–28.

Kurashima, A. 1958a: Summer monsoons. *Kagaku* **28**, 393–400 (in Japanese).

Kurashima, A. 1958b: Monsoon I–III. Tenmon to Kisho. **24**(5) 7–10, (6) 2–6, (7) 8–14 (In Japanese).

Kurashima, A. 1959a: General circulation and monsoon. *In* Kisetsufuu (monsoon), Chijinshokan, Tokyo, 201–283 (in Japanese).

Kurashima, A. 1959b: Classification of monsoon and its geographical distribution. *In* Kisetsufu (monsoon), Chijinshokan Tokyo 14–28 (in Japanese).

Kurashima, A. 1968: Studies on the winter and summer monsoons in East Asia based on dynamic concept. *Geoph. Mag.* **34** 145–235.

Murakami, T. 1958: The sudden change of upper westerlies near the Tibetan Plateau at the beginning of summer season. *J. Met. Soc. Japan* **36** 239–247 (in Japanese).

Murakami, T. 1959: The general circulation and water-vapour balance over the Far East during the rainy season. *Geoph. Mag.* **29** 131–171.

Otani, T. 1954: Converging line of the northeast trade wind and converging belt of the tropical air current. *Geoph. Mag.* **25** 1–122.

Raman, C. R. V., Dixit, C. M. 1964: Analysis of monthly mean resultant winds for standard pressure levels over the Indian Ocean and adjoining continental areas. *In* Preceedings of the symposium on tropical meteorology, Rotorua, New Zealand, 107–118.

Saito, N. 1966: A preliminary study of the summer monsoon of southern and eastern Asia. *J. Met. Soc. Japan* **44** 44–59.

Academia Sinica, Staff Members, Peking, 1957, 1958: On the general circulation over Eastern Asia (1), (2), (3). *Tellus* **9** 432–446, **10** 58–75, **10** 299–312.

Stremousov, N. V. 1935: K voprosu o sinopticheskikh protsuessakh vostochnoi chasti aziatskogo materika i prilegayushchikh morei. *J. Geofiz.* **2** 204–221.

Sugimoto, Y. 1967: Okhotsk High in Bai-u season. *Japan Met. Agency* 1–28 (in Japanese).

Tao, Shih-yen, 1948: The mean surface air circulation over China. *In* Collected Scientific Papers, Academia Sinica 1954 575–584.

Thompson, B. W. 1951: An essay on the general circulation of the atmosphere over Southeast Asia and the West Pacific. *Quart. J. Roy. Met. Soc.* **77** 569–597.

Tu, Chang-Wang, Hwang, Sze-Sung 1944: The advance and retreat of the summer monsoon in China. *In* Collected Scientific Papers, Academia Sinica 1954 519–534.

Yao, Chen-Sheng, 1939: The stationary cold fronts of central China and the wave-disturbance developed over the Lake-Basin. *In* Collected Scientific Papers, Academia Sinica 1954 203–214.

Yoshino, M. M. 1963: Rainfall, frontal zones and jet stream in early summer over East Asia. *Bonner Met. Abh.* (3) 1–127.

Yoshino, M. M. 1965: Four stages of the rainy season in early summer over East Asia (Pt. I). *J. Met. Soc. Japan II* **43** 231–245.

PRECIPITATION DISTRIBUTION AND MONSOON CIRCULATION OVER SOUTH, SOUTHEAST, AND EAST ASIA IN SUMMER

Masatoshi M. Yoshino and Haruhiko Aihara

Abstract: An attempt was made to clarify the distribution of monthly precipitation in summer in connection with the monsoon circulation patterns over South, Southeast, and East Asia. The distribution maps of the precipitation were made using the available data for June, July, and August, 1959–1968. The distribution patterns of precipitation were classified into four types, Ia, Ib, IIa, and IIb, and composite maps for each type were presented. In addition, composite maps of the pressure distribution at the surface level were also made for each type. In comparing the maps of precipitation with the maps of pressure for the corresponding type, it was pointed out that the difference in the distribution of precipitation is ascribed to the pressure distribution, especially intensity, shape and position of the anticyclone in the Northwest Pacific, the low pressure region over South Asia and the low pressure belt in the tropical Pacific. Types Ia and IIa appeared mostly in the early summer months: the low pressure belt in the tropical Pacific locates along the equator and the clear, but rather narrow precipitation zones run along the northern and the southern margins of the low pressure belt. In contrast, Types Ib and IIb appeared in the late summer month: the western part of the low pressure belt in the tropical Pacific goes up north, to the Philippines in association with the retreat of the anticyclone over the Northwest Pacific, and the northern precipitation zone mentioned above becomes wider due to the larger amount. Similar close relationships between pressure and precipitation distribution were found also in other parts of Asia in the maps presented.

1. Introduction

The term "Monsoon Asia" implies a region experiencing monsoon conditions of climate as stated by Robinson (1966). Monsoon means a seasonal wind, but "monsoon season" is used to mean wet season. The term is used by people in Southeast Asia to mean simply rainfall. As a result of the varied uses of the term monsoon, geographers and climatologists in other regions of the world often erroneously understand that Monsoon Asia is a wet region with abundant rainfall over the entire area.

Of course, it can be said that Monsoon Asia as a whole is wetter than other regions of the Northern Hemisphere, as revealed by the study on precipitable water in Monsoon Asia by Asakura (1971). However, there are dry regions even in South and Southeast Asia. In other words, it may be said that the humid regions are restricted.

Generally, it is known that the rainfall in Monsoon Asia is characterized by its seasonality, interannual variability and torrential nature. However, there exists another characteristic feature: i.e. its regionality. We intend, therefore, to make clear the regionality of the rainfall in Monsoon Asia, considering the problem mentioned above. Regions of annual rainfall deficiency in the Indian subcontinent, in tropical Southeast Asia and in East Asia were discussed in detail by Trewartha (1962). In the present study, the regional differentiation of rainfall in South, Southeast, and East Asia is investigated in relation to pressure patterns as causes of the monsoon circulation from the viewpoint of synoptic climatology.

2. Previous Studies on the Rainfall Distribution

At the end of the last century, Supan (1898) dealt thoroughly with the distribution of precipitation over the continents. Since then, through climatology, new light has been shed on the amount of precipitation as well as its duration in Monsoon Asia, considering the paths of cyclones and winds as causative factors on the one hand and vegetation as one of the formative factors for landscape on the other (Passerat, 1906). There was some uncertainty in his paper as Woeikov, one of the earliest climatologists who completed a study of world climatography (Woeikov, 1887), pointed out (Woeikov, 1907), but Passerat's description was detailed and dynamic enough for the time.

Kendrew (1927) had already described the regionality of the monsoon rains in his comprehensive study on climatography. For instance, he pointed out that the wet strip on the west coast of India and Ceylon, which has enormous downpours of 1,250–2,500 mm from June to September, does not extend far north of Bombay. Similarly, east of the Bay of Bengal, a strikingly rapid transition from the rainy coast and windward slopes to the mountainous regions in the Arakan Range can be found. Beyond the range, a dry region appears in central Burma. Such a description on the mean state of rainfall in association with air flow conditions was made for all of Asia.

In monographs dealing with the regional geography of Monsoon Asia, rainfall distribution with maps have also been mentioned, e.g., a map of annual rainfall on a scale of 1:55,000,000 (Sion, 1928). During World War II, some Japanese climatologists, Arakawa (1942), Fukui (1942) and Ogasawara (1945), contributed to studies on the rainfall regime in Southeast Asia. Notably, Ogasawara presented detailed maps of monthly rainfall distribution and described the distribution in relation to airflows from a dynamic standpoint. The year by year variation in distribution caused by different regional airflow developments was, however, not treated.

Since World War II, many books have been published on the regional geography of Monsoon Asia, e.g., Dobby (1958), Ginsburg (1959), Stamp (1962), Cressey (1963), Fisher (1965), Robinson (1966), and Johnson (1969). Because these books were primarily intended to describe social, economic or political situations, they dealt with rainfall distribution in only a few pages, even though the descriptions have become more concrete with the knowledge of modern synoptic and aerological climatology. For instance, concerning rainfall distribution for the months of June to August on the Burmese coast, Fisher (1965) wrote that the belt of heaviest precipitation is not unduly wide, and that its southern edge seems to correspond quite closely to the mean position of the jet axis.

Rainfall variability in relatively restricted regions has been stressed. An example from a region south of Ahmadabad in Gujerat, India, was presented by Spate (1957): the region has a mean annual rainfall of 1,600 mm, with a minimum of 1,000 mm and a maximum of

2,900mm. If it rains regularly, it is known that paddy rice in this region can be cultivated even with a 1,100 mm annual rainfall. Much attention has been given to rainfall variation or range of annual rainfall since the end of the 19th century (Hann, 1910), because rice culture depends largely on the amount of rainfall. The study of variability of rainfall continued after World War II, e.g., in India (Chatterjee, 1953), in eastern China (Yao, 1958 and 1965) and in Japan (Daigo, 1947). These studies revealed that the distribution of the variability of annual or monthly precipitation does not coincide with the distribution of annual or monthly precipitation and that its value changes from season to season. This may support the idea of area variations from one year to another, that is, annual distribution differences must be studied.

On the other hand, the development of aerological networks after World War II has made it possible to detect the rainfall patterns associated with the tropospheric circulation characteristics in this region. For instance, the water-vapor transport over East Asia is directed from west to east and only south of about 30°N from southwest to northeast (Flohn and Oeckel, 1956). This result contrasts with the usual geographers' concept that the summer rains are produced by a large scale inflow of moist maritime air to the continent in Monsoon Asia (Flohn, 1960). After analyzing the upper air data collected at about 125 stations over East Asia, Murakami (1959) concluded that in the early stage, of the rainy season, the moisture flow controlled by the westerlies from India definitely influenced the distribution of rainfall. In the last stage, however, the moisture flow originating from the Pacific and taking a roundabout route over south China and the East China Sea becomes more significant. Recently, by analyzing the distributions of the water vapor and wind system over South and East Asia, the structure of the moist air inflow from Southeast Asia to East Asia was made clear by Saito (1966). The wettest region was usually formed to the north of the monsoonal inflow rather than in the monsoon itself in East Asia. The rainfall together with the frontal zones and jet streams in early summer over East Asia were studied by Yoshino (1963) and the change in rainfall distribution during the Baiu according to four stages was presented (Yoshino, 1965 and 1966).

The relationships between the rainfall and the high level anticyclone or the tropical easterly jet over South Asia should be mentioned, although such relationships have not been taken into consideration in the present study. As Ramanathan (1960) pointed out, the rainfall which depends on the moist air inflow of the monsoon winds at lower levels and their ascent is an important agent in changing the position and strength of the warm subtropical anticyclone at high levels to the north of the monsoon area, and correspondingly of the high level easterly winds over South Asia, and indirectly in guiding the movement of monsoon depressions and storms. Flohn (1964) presented the fact that a Hadley circulation is observed in the entrance region of the tropical easterly jet over Southeast Asia, inducing large-scale convection.

Recent reviews on rainfall distribution can be found in climatographic studies of Asia such as those edited by Hatakeyama (1964) and Arakawa (1969).

Using the results of two years (1965 and 1966) of weather-satellite cloud observations, the average monthly cloudiness in the tropics was presented for each month (Sadler, 1969). Sadler compared cloud distribution with mean rainfall distribution over Africa and came to the conclusion that, despite the disparity of sample years, there was a crude agreement between their patterns. If such a relationship could be established for Southeast Asia and the oceans, where the rainfall data are insufficient, a distribution map of cloudiness would provide useful information on detailed rainfall distribution. Secondly, comparing the cloud distribution for two different years, it was pointed out that the maximum areas were cloudier in 1965 than in 1966, especially over the western tropical Pacific in February, June, and July. The same tendency was found in South Asia. Thus, the inter-annual variation within the maximum areas was large. This suggests that the inter-annual variation of rainfall is equally large along the intertropical convergence zones.

3. DATA ARRANGEMENT

The rainfall data at about 170 stations were taken from the "World Weather Record," "Monthly Climatic Data for the World" and the original data available at the Long-range Weather Forecasting Division of the Japan Meteorological Agency. We made distribution maps of the monthly rainfall for June, July, and August, 1959–1968. For China, only data for July, 1959, were obtained from a figure showing anomalies (%) of precipitation presented by Shen (1965) and for August 1961 and 1962, the monthly precipitation amount by Kao and Yang (1965). The maps were then classified, with special attention paid to the following characteristics in the distribution patterns: a) The distribution of heavy rainfall areas in northeastern India, East Pakistan, and western Burma. b) The rainfall zone running from southern China to western Japan. c) The double rainfall zones in the region over Korea and Japan. d) The shape of the rainfall zone extending parallel to 5°–10°N over the equatorial Pacific. e) Whether the heavy rainfall area over the Philippines belongs to the rainfall zone over East Asia or over the equatorial Pacific. f) The shape, mainly the length, of the rainfall zone extending from eastern New Guinea to the Moluccas.

On this basis, we classified the maps into four types:

Type Ia—August 1959, June 1960, June 1961, July 1962, and June 1964.

Type Ib—August 1961, July 1963, August 1961, July 1968, and August 1968.

Type IIa—July 1959, June 1962, June 1965, June 1966, June 1967, and June 1968.

Type IIb—August 1960, July 1961, August 1962, August 1964, and August 1966.

The rest of the maps were omitted in the classification because of indefinite distribution patterns. Then composite maps of each type were made by calculating the mean values at 338 points on the maps. The results are given in Figs. 1–4.

Composite maps of pressure distribution at the surface level were also made, using 252 points, 5° × 5° grid points, for each type. The pressure data were taken from the "Witterung in Übersee" published by Deutscher Wetterdienst, Seewetteramt in Hamburg. The composite maps are shown in Figs. 5–8.

4. RESULTS AND SOME CONSIDERATIONS

Type Ia

Type Ia appeared in the early summer months except in August, 1959. From the rainfall distribution given in Fig. 1, the following should be noted: a) In western Ghats, the west coast of the Indian subcontinent, and in the coastal region of southeastern East Pakistan and of Arakan in Burma there was heavy rainfall. b) West Pakistan has less than 50 mm, but more than 0 mm. c) The heavist rainfall regions in the eastern part of the Indian subcontinent appear in eastern Bihar and in northwestern East Pakistan, but not in Assam. d) The regions from the coast in southeastern Thailand to inner Cambodia as well as northern Laos receive relatively much rainfall. e) A single rainfall zone develops from east of Formosa to the southern coast of Japan. f) The rainfall zone runs due parallel to a latitude between 5°

and 10°N over the equatorial Pacific. g) The rainfall zone develops from eastern New Guinea to the Moluccas.

The pressure distribution of Type Ia shown in Fig. 5 revealed the following: a) A low centered in West Pakistan has 996 mb and the region lower than 1,004 mb occupies a broad area of the continental part of East Asia. b) The anticyclone over the northwest Pacific is weak in the area shown in Fig. 5. Only a small region shows over 1,016 mb. c) Between these, the isobars run in a relatively south-north direction over East Asia as compared with the other types. d) Over the equatorial Pacific, the eastern part (east of 165°E) of the low pressure belt locates in the Southern Hemisphere and the western part runs along 5°N. The lowest part (1,008.4 mb) appeared in the region between 145° and 155°E of 5°N.

Comparing Fig. 5 with Fig. 1, it should be pointed out that the following characteristics of rainfall distribution are closely related to the pressure pattern. a) The heaviest rainfall region in the northeastern Indian subcontinent appears west of 90°E, when the trough along 85°E over the Bay of Bengal develops locally but clearly. The same relation is found in Type IIb (Figs. 4 and 8). If this trough does not develop, as in Type Ib, the heaviest rainfall region locates east of 90°E. b) A relatively weak anticyclone over the northwest Pacific and the running direction of the isobars over the Philippine Sea and the East China Sea caused a markedly developed rainfall zone from the Philippines to the southern coast of Japan via the Ryukyus. This rainfall zone can be traced further east along 35°N over the Pacific, being followed by the Pacific polar frontal zone. It has been said that the Taiwan convergence has little rainfall activity (Thompson, 1951; Watts, 1955; Yoshino, 1963), but striking activity is found along the convergence in the case of Type Ia. c) The rainfall zone over the equatorial Pacific locates at the northern margin (between 5° and 10°N) of a low pressure belt whose axis runs along 5°N. The rainfall zone corresponds to the northern intertropical convergence zone over this region illustrated by Yoshino (1969). Another rainfall zone over New Guinea to the Moluccas and Celebes is considered to be a consequence of strong convective activity along the southern intertropical convergence zone, also illustrated by Yoshino (1969) or page 98 of this volume.

Type Ib

With respect to the rainfall distribution of Type Ib (Fig. 2), we can point out the following: a) Western Ghats also has heavy rainfall, but, in contrast, a scanty rainfall zone appears from northwestern Ceylon to Aurangabad via Bangalore and Bellary. b) West Pakistan has less than 50 mm, but more than 0 mm just as in Type Ia. c) The heaviest rainfall region in the eastern part of the Indian subcontinent appears in Assam, especially on the Khasi Hills near Cherrapunji. d) From this region, a region with relatively greater rainfall of over 300 mm extends westward in the Ganges valley. Delhi is situated roughly on its border. e) Double rainfall zones develop over East Asia: one along the southern coast of Japan and another from northern Korea to the Hokuriku District, Japan. These double zones were mentioned in previous papers (Yoshino, 1963 and 1966) as observed at the later stage of the rainy season. f) The rainfall zone over the equatorial Pacific widens and, by extending north at its western part, continues to the heavy rainfall region in the Philippines and further to Hong Kong. g) In contrast, the rainfall zone south of the equator occupies only the eastern part of New Guinea. h) Rainfall lower than 100 mm is observed in the southern equatorial Pacific east of 160°E.

With respect to the pressure distribution of Type Ib shown in Fig. 6, the following phenomena must be noted: a) The intensity and shape of the low over the continent is almost the

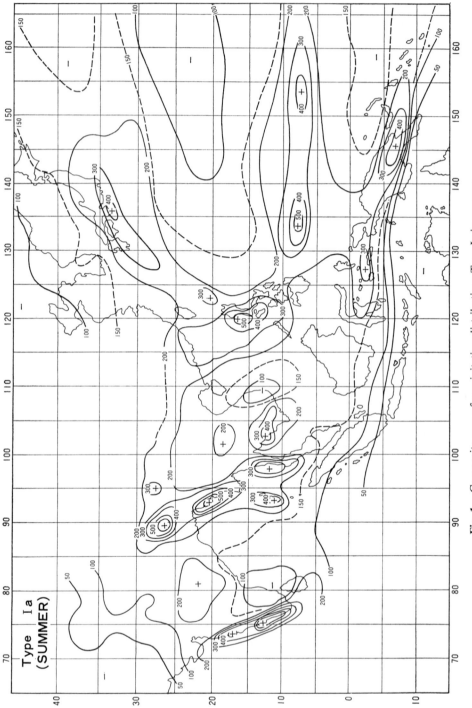

Fig. 1. Composite map of precipitation distribution; Type Ia in summer.

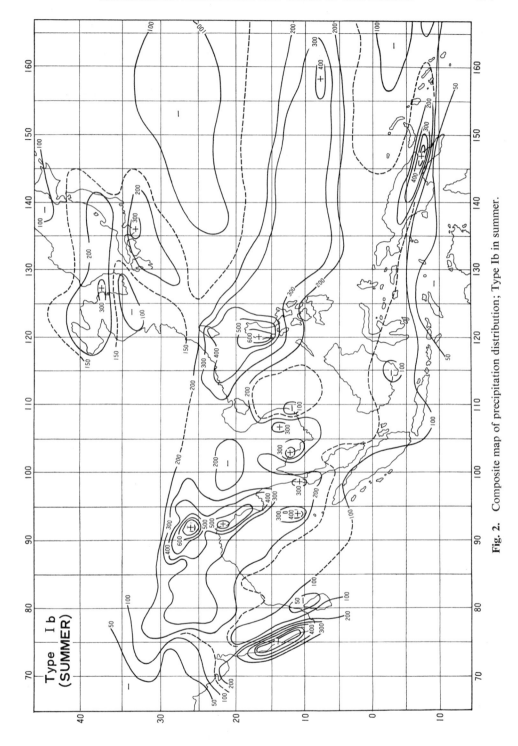

Fig. 2. Composite map of precipitation distribution; Type Ib in summer.

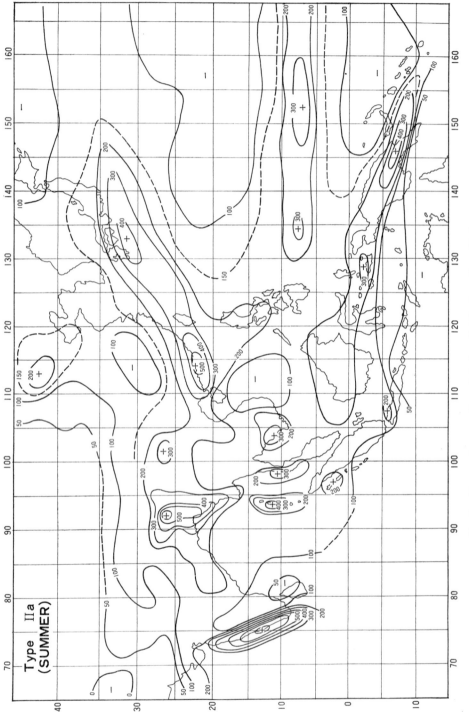

Fig. 3. Composite map of precipitation distribution; Type IIa in summer.

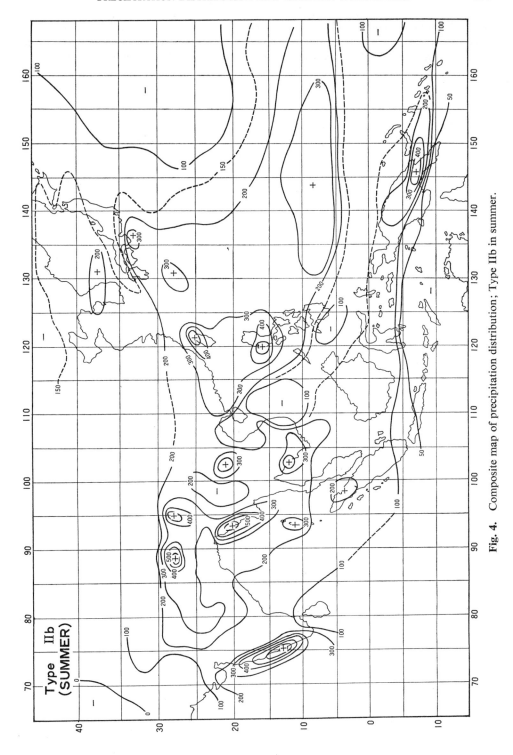

Fig. 4. Composite map of precipitation distribution; Type IIb in summer.

same as Type Ia. b) The local trough along 85°E over the Bay of Bengal is weaker in Type Ib than in Type Ia. c) The anticyclone over the Northwest Pacific is stronger, and its center shifts northward. d) Accordingly, the pressure gradient parallel to 30°N over the Northwest Pacific is greater than in the case of Type Ia. e) In contrast, the pressure gradient east-west over the South China Sea and the Philippine Sea is weaker. f) The low pressure zone starts from the equatorial Pacific and goes north to the Philippine Sea. Its center is 1,006.8 mb in the region between 150°–155°E of 5°N, and has the lowest value among the four types. g) The region around the Celebes Sea, Celebes and the Moluccas is higher in Type Ib than in other types. h) The pressure gradient over Borneo, Sumatra, and Java is greater.

Comparing Fig. 6 with Fig. 2, the following observations can be made: a) The resaon why the heaviest rainfall region appears in Assam in Type Ib could be ascribed to a weaker trough over the Bay of Bengal, which induces a moist airflow from the Bay of Bengal to orographic rainfall on the Khasi Hills near Cherrapunji. b) The rainfall zone over the equatorial Pacific appears to coincide with the low pressure belt. But, strictly speaking, the rainfall zone is located north of the low pressure belt east of 150°–155°E. At the western part, their axes are situated in the same position. c) Because of the relatively high pressure over Celebes and the Moluccas, less rainfall is observed as compared with Type Ia. d) Due to the greater pressure gradient indicating stronger monsoon circulation, the rainfall over Sumatra and Java in Type Ib is more than in Type Ia.

As indicated earlier, Type Ib was established on the basis of data obtained during two months of July and three months of August. Therefore, it can be said that Type Ib appears frequently in late summer while Type Ia appears in early summer. In late summer, the trough along 85°E over the Bay of Bengal is apparently weakened, because of a lowering of pressure in the Ganges valley. Also, the low pressure belt over the equatorial Pacific goes up north to the Philippine Sea at its western part, which is one of the main paths of typhoons. The change between the early and late summer months in these pressure patterns in association with monsoonal circulation causes the difference in rainfall patterns observed in Types Ia and Ib.

Type IIa

The rainfall distribution of this type is given in Fig. 3. Its characteristics are the following: a) West Pakistan has a small amount of rainfall. Quetta and its northern mountaneous region, in particular, has O mm. b) The heaviest rainfall region in the eastern part of the Indian subcontinent is located in the region surrounding the Khasi Hills in Assam and the border of East Pakistan and Assam. c) The rainfall zone extending from Hong Kong to the southern coast of Japan is apparent. d) The Philippines have less rainfall than the other types. e) Rainfall in the lower Yangtze region, China, is less than normal. f) In inner Mongolia and western Manchuria, the rainfall is over 200 mm. g) Over the equatorial Pacific, the rainfall zone develops between 5° and 10°N. From eastern New Guinea to the Moluccas conditions are the same as in Type Ia. h) In the remaining regions the rainfall is less than 100 mm. In other words, a rather dry condition develops in broad areas over the Pacific as compared with Types Ia and Ib. i) Some parts of Sumatra, Borneo and Java are wetter.

The following observations can be made with respect to the pressure distribution in Fig. 7: a) The heat low in West Pakistan is 996 mb. Its center (994.3 mb) is the lowest of the four types. b) On the other hand, the pressure is higher over East and Southeast Asia. The isobar 1,006 mb runs across Thailand and the isobar 1,010 mb runs over the Philippines, the Celebes Sea and Borneo. Accordingly, the pressure gradient is greater over the Tibetan Plateau and smaller over the South China Sea and eastern China. c) A lower pressure region is found in inner Mongolia and Manchuria. d) The anticyclone over the northwest Pacific is stronger, i.e. the isobar 1,012 mb envelops a broad area with its center at 1,016 mb. e) A low pressure belt occupies the tropical region with its axis along the equator. Its lowest center is, however, as low as 1,009.2 mb. This is the highest value in the tropical low pressure belt among the four types. f) Over the Southern Hemisphere, the tendency is almost the same.

Lastly, the relationships between rainfall and pressure distribution patterns in Type IIa to be noted are the following: a) In spite of the clearly developed trough running from 80°E, 20°N, to 85°E, 15°N, and then along 85°E to south over the Bay of Bengal, the rainfall over the Indian subcontinent is less, probably due to the weaker cyclone activity in this type, as compared to the early summer months. b) The relatively rapid increase of pressure in eastern Tibet, which means a weaker inflow of southwest air streams over eastern China, might cause anomalously low rainfall over eastern China. For instance, in July 1959, as noted by Shen (1965), the rainfall anomaly was -50% in the Kiang Hwai region, central East China, and the minus departure of water vapor transport was -50% in the region of the lowest course of the Yangtze river as compared to the condition in July 1957, a normal year. c) The rainfall zone extending from the coast of southern China to western Japan corresponds to the direction of southwesterly air flow from the South China Sea to the Ryukyus, which can be easily estimated by the running direction of the 1,008 mb isobar. d) The low pressure area extends to inner Mongolia, where a rainfall of more than 200 mm is observed. e) Areas with a rainfall of less than 100 mm are broad, because a relatively higher pressure predominates over the Pacific. f) The rainfall zones, one running parallel to the latitude between 5° and 10°N and another from eastern New Guinea to the Moluccas, are situated just at the margin of the low pressure belt, which is expressed by the 1,010 mb isobar over the equatorial Pacific. g) Over the Philippines, heavy rainfall is not observed, because the pressure there is higher than the South China Sea and the equatorial Pacific.

Type IIb

In Fig. 4, the following rainfall distribution characteristics of Type IIb are especially to be noted: a) In western Ghats and in the coastal region of the southeastern part of East Pakistan, there is also found a strip region with heavy rainfall, as with the other types. b) West Pakistan has little rainfall, as in the case of Type IIa. c) The heavy rainfall in the eastern part of the Indian subcontinent does not appear in the Khasi Hills region, but in two other regions: in Dibrungarh in the Brahmaputra valley, northern Assam, and in Darjeeling in Sikkim and its surrounding regions. d) A higher rainfall region is found in the Ganges valley, including these regions, just as in Type Ib. e) In Ceylon, no marked contrast is found between the southwest and northeast regions. f) There is also no heavy rainfall strip zone on the southernmost Andaman coast of Burma. g) Over Korea and Japan, double rainfall zones can be observed, but not more clearly than in Type Ib. h) The broad rainfall zone from the equatorial Pacific via the Philippine Sea to southwestern Japan is most striking in this type. i) A dry zone is observed east of 140°E along the equator, which gradually goes up north into the Celebes Sea, and is connected with the dry area in the South China Sea. j) The rainfall zone over New Guinea is smaller, as in Type Ib.

Concerning the pressure distribution, the following points should be noted: a) A low is found over the Asian continent and along 85°E over the Bay of Bengal. Its shape is almost the same as in Type Ia. b) The anticyclone over the Northwest Pacific is strong, but its center is shifted slightly north. c) The broad low pressure area starts from the equatorial Pacific and

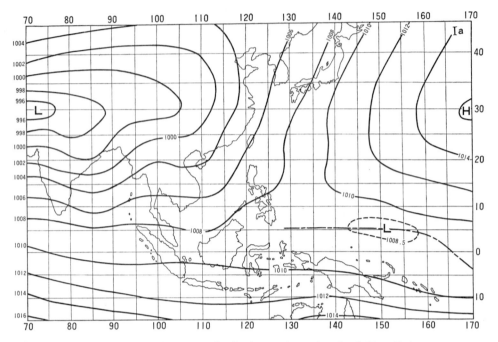

Fig. 5. Composite map of pressure distribution at the surface level; Type Ia in summer.

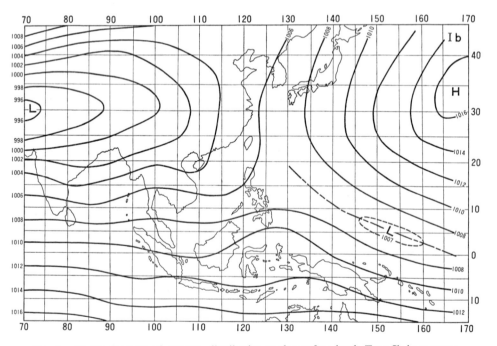

Fig. 6. Composite map of pressure distribution at the surface level; Type Ib in summer.

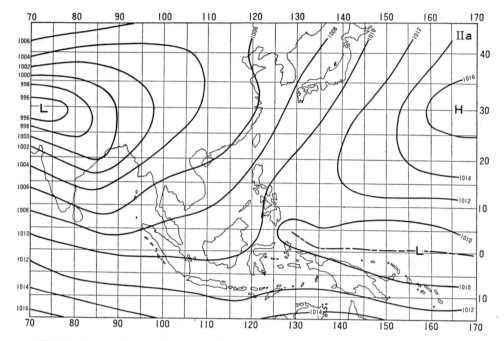

Fig. 7. Composite map of pressure distribution at the surface level; Type IIa in summer.

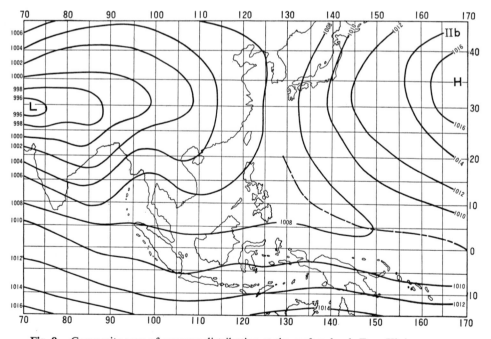

Fig. 8. Composite map of pressure distribution at the surface level; Type IIb in summer.

extends to the Philippine Sea. d) Over the Southern Hemisphere, south of 15°S in particular, the pressure rises rapidly. This tendency is found in Types Ib and IIb.

The relationship between the rainfall pattern and the pressure pattern is as follows: a) The heavy rainfall center west of 90°E over the northeastern Indian subcontinent can be explained as in the case of Type Ia in Fig. 1. b) The relatively developed ridge over the Andaman Sea and the Gulf of Siam may be caused by the weak southwest flow, which results in low orographic rainfall over Ceylon and the southern Andaman coast of Burma. c) The broad rainfall zone over the equatorial Pacific via the Philippines, Taiwan and the Ryukyus to southwestern Japan, corresponds to a low pressure area through which typhoons pass frequently during these months. d) The low rainfall area over the Southern Hemisphere goes up to 5°S, around the southern Celebes, due to the higher pressure over northern Australia. e) The anticyclone causes a large low rainfall region over the Northwest Pacific.

5. SOME CONSIDERATIONS

a. *Rainfall in the Western Indian Subcontinent*

Northward from the Gulf of Cambay, India, the rainfall decreases rapidly. This can be explained by two factors (Miller, 1959): 1) the trajectory of the air currents, and 2) the dryness of the air over Sind. Miller also added that an east–west line drawn near Karachi (roughly 25°N on the coast) divides those winds to the south which have travelled over the Arabian Sea from those to the north which have travelled over the arid lands of Arabia and Baluchistan, and this line practically marks the limit of monsoon rainfall. As was shown in Figs. 1–4, this limit is expressed roughly by the 50 mm isohyet in each type. The rainfall in southeastern Afghanistan and its adjacent region in West Pakistan is actually zero in Types IIa and IIb, against the evidence that the rainfall is quite small, but is not zero in Types Ia and Ib. One reason for this is that in the case of Types IIa and IIb, the airflow has a stronger western component, because the isobars run from WNW to ESE over Baluchistan. On the other hand, the isobars run W to E, indicating a more southerly component revealing the influence of the Arabian Sea upon the airflow there.

A comparison of maximum rainfall latitudes along the west coast with the trough axes in the mid-troposphere during active monsoon periods over the Arabian Sea reveals that there is a high correlation ($r = 0.95$) (Miller and Keshavamurthy, 1968). But in early summer, the heaviest rains occur almost directly below the mid-tropospheric trough and to the south of 15°N. On the other hand, in late summer, the maximum rainfall occurs 100–150 km south of the axes of the mid-tropospheric troughs. In the present study, it was shown that the maximum rainfall in western Ghats appears not only south of 15°N, but also north of 15°N in Types Ia and Ib, and, only south of 15°N in Types IIa and IIb. Therefore, it can be said that the axes of the mid-tropospheric trough run at a relatively higher latitude in Type Ia and Ib than in Types IIa and IIb. Furthermore, the evidence that a zero rainfall region appeared in southeastern Afghanistan and its adjacent region in West Pakistan in Types IIa and IIb must also be related to the location of the axes of the mid-trospheric trough which run at a relatively lower latitude in these types.

Ramage (1966) concluded that large-scale subsidence causes the droughts in parts of the Arabian Sea and Northwest India. Desai (1967), on the other hand, refused to accept Ramage's model and his various interpretations and suggested that inversion over these regions is due to airmass properties. Subsequently, Flohn (1968) discussed the problem, giving the patterns of surface divergence and vertical velocity. He concluded that it seems unlikely that the 500 km wide convergence zone off the west coast of the Indian subcontinent is related to orography alone.

In the present study, the wind field dynamics and development of heat lows have not been considered, but only the difference in the composite patterns of the various types is emphasized.

b. *Rainfall in New Guinea and Its Vicinity*

The rainfall data for the tropical regions are quite insufficient for the present study. According to a study by Brookfield and Hart (1966), there are 937 stations in New Guinea and its adjacent regions, if we utilize every rain gauge with one complete year or more of records. As the data for 1960 is made available in their report, rainfall distribution maps for New Guinea and the Solomons in June 1960 and August 1960 are presented here. As mentioned earlier, June 1960 belongs to Type Ia and August 1960 to Type IIb. Therefore, a comparison can be made insofar as the rainfall patterns in these two types are concerned.

The distributions given in Figs. 9 and 10 show very complex features as compared to Figs. 1–4. Roughly speaking, however, the distribution pattern in June 1960 is fairly close

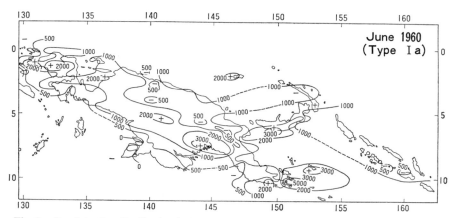

Fig. 9. Precipitation distribution in New Guinea in the case of Type Ia in June 1960.

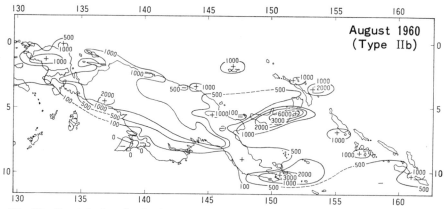

Fig. 10. Precipitation distribution in New Guinea in the case of Type IIb in August 1960.

to that in August 1960 on the following points: i) The region including the southern side of the central cordillera, the western Hollandio districts and the Geelvink Bay region, West Irian, has a rainfall of over 1,000 mm. ii) The southern side of the range in the Vogelkop Peninsula, the westernmost part of New Guinea, shows more than 1,000 mm. iii) The region from New Britain to the Huon Peninsula frequently has rainfall over 2,000 mm and locally over 3,000 mm. iv) Milne Bay, the southeastern part of New Guinea, also has much rainfall, and locally over 3,000mm. v) In contrast to these regions with much rainfall, there are regions with zero rainfall in the area in and around Frederik Hendrik Island. vi) The coastal areas facing the Arafura Sea generally have a small amount of rainfall. vii) The mountainous region in the main central range of the Territory of New Guinea shows a relatively lower rainfall (200–800 mm), which is unexpected.

The following differences can be observed between Figs. 9 and 10: i) Heavy rainfall of over 2,000 mm appeared locally at many points in June 1960, but was less widespread in August 1960. ii) As stated above, the region from New Britain to the Huon Peninsula in eastern New Guinea, and the Milne Bay region in southeastern New Guinea had heavy rainfall. An extraordinarily heavy rainfall of over 6,000 mm appeared in the latter region in June, however, and in the former in August. For instance, the maximum, 6,177 mm, was recorded at Sewa Bay Station (10°02′S, 150°58′E, 50 ft), southeastern New Guinea, in June 1960, while a maximum of 6,242 mm was reported at Pomio Station (5°20′S, 151°36′E, 350 ft), New Britain, in August 1960.

This difference may be interpreted as a generally observable difference between Types Ia and IIb. As can be seen in Figs. 5 and 8 or in the maps of the stream lines and convergence zone of the mean surface winds for June and August presented by Sandoval (1967), the NITC is located further north in August than in June. Accordingly, the SITC is located roughly along the equator in August, but at about 5°S in June. In other words, it can be said that the SITC which extends over New Guinea and ESE is more apparent in June, inducing stronger ESE monsoons over the Southwest Pacific. This may be the cause of the difference.

A sharp gradient from 2,000 mm to 0 mm at a distance of 300–400 km on the southern side of the central cordillera of New Guinea should be noted. Topographical effects such as differences between the windward and leeward of the mountain range, the height of the central cordillera and the heat effect of an isolated island should be observable. This was confirmed recently by the masses of heavy cloud seen on satellite pictures. It remains, however, a problem why 0 mm of rainfall were recorded in June 1960 on the southern coast of New Guinea, for instance at Frederik Hendrik Island facing the Arafura Sea and some other places. Here, emphasis is placed on the role played by the seasonal change in the convergence zone on the rainfall distribution over the central cordillera of New Guinea.

c. *The Rainfall Distribution Map of Monsoon Asia: Three examples* (Figs. 11–13)

Distribution maps of the monthly precipitation were made for three months: August 1961 as an example of Type Ib, July 1959 for Type IIa and August 1962 for Type IIb. These months were chosen because distribution maps of rainfall in China are available for August 1961 and 1962 in the figures presented by Kuo and Yang (1965) and for July 1959 by Shen (1965).

First, it should be stated again that the data for the tropical region is insufficient. It is clear that the isohyets are simpler in the Southwest Pacific because of scarce observation points. If we compare the distribution characteristics in New Guinea in Figs. 11–13 with those in Figs. 9 and 10, we can easily understand how serious this problem is. The representativeness

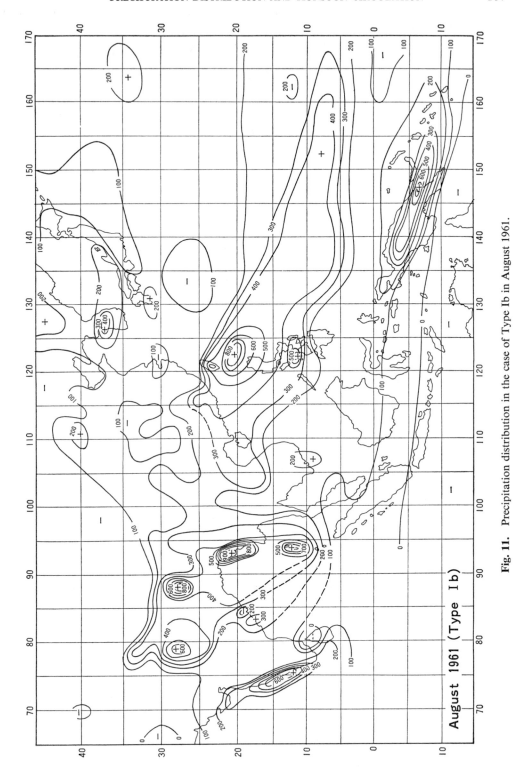

Fig. 11. Precipitation distribution in the case of Type Ib in August 1961.

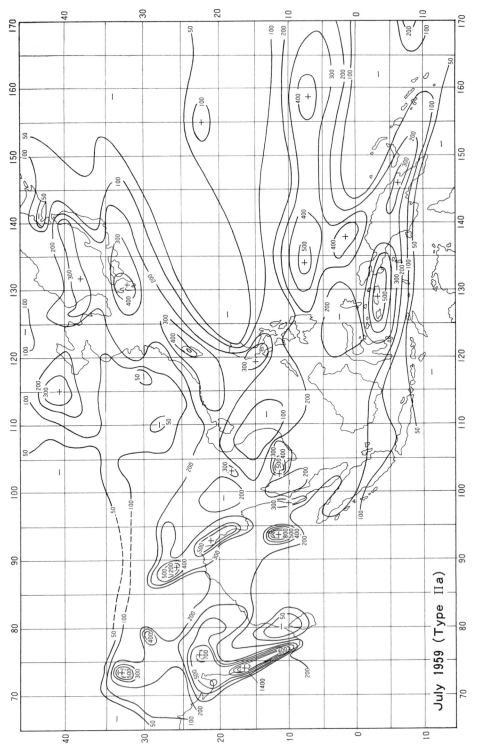

July 1959 (Type IIa)

Fig. 12. Precipitation distribution in the case of Type IIa in July 1959.

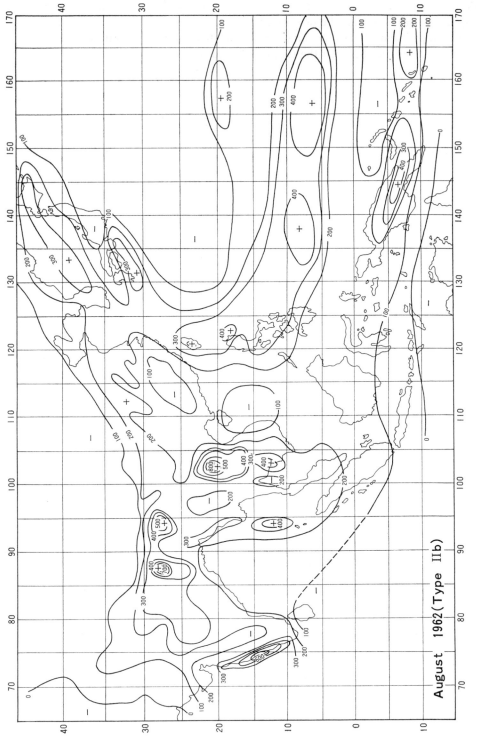

Fig. 13. Precipitation distribution in the case of Type IIb in August 1962.

of the observed rainfall at stations in these tropical region, where the local convective activity is very striking, must be taken into consideration.

In spite of problems due to the data used, the distribution patterns do show significantly different characteristics between the types described in Section 4.

6. Summary

This paper presents first the results of previous studies on rainfall distribution and its relation to circulation patterns, water vapor transport and cloudiness distribution as revealed recently by satellite pictures. On the basis of these results, monthly rainfall distribution maps for June, July, and August, 1959–1968, were classified into four types. Then composite maps of each type were drawn for rainfall distribution and pressure at sea level.

The rainfall distribution is closely related to the pressure distribution in the respective types. For instance, it was shown that the trough along 85°E over the Bay of Bengal, the anticyclone over the Northwest Pacific, the low pressure belt with its axis (NITC) along 5°N, and a strong convective activity along the SITC running over New Guinea to the Moluccas strikingly conditioned the distribution of precipitation in Monsoon Asia. The first two determine the behavior of the so-called SW monsoon in South Asia and the polar frontal zones in East Asia and, accordingly, the distribution of precipitation.

Lastly, rainfall in the western Indian subcontinent and over New Guinea were dealt with in detail. Examples of the monthly rainfall distribution in Monsoon Asia for July 1959, August 1961, and August 1962 were presented, to make clear the distribution of the fundamental climatic conditions of moistness and dryness in the region.

References

Arakawa, H. 1942: Daitôa no kikô (Climate of Southeast Asia). Tokyo 199p.
Arakawa, H. 1969: Climates of Northern and Eastern Asia. Amsterdam 248p.
Asakura, T. 1971: Transport and source of water vapor in the northern hemisphere and Monsoon Asia (*In* this volume).
Chatterjee, S. B. 1953: Indian climatology. Calcutta 417p.
Cressey, G. B. 1963: Asia's lands and peoples. New York 663p.
Daigo, Y. 1947: Handy atlas of agricultural meteorology in Japan. Tokyo 234p.
Desai, B. N. 1967: The summer atmospheric circulation over the Arabian Sea. *Jour. Atmos. Sci.* **24**(2) 216–220.
Dobby, E. H. G. 1958: Southeast Asia. London 415p.
Fisher, C. A. 1965: Southeast Asia. London 821p.
Flohn, H., Oeckel, H. 1956: Water vapour flux during the summer rains over Japan and Korea. *Geoph. Mag.* **27** 527–532.
Flohn, H. 1960: Recent investigations on the mechanism of the "summer monsoon" of Southern and Eastern Asia. *In* Monsoons of the World 75–88.
Flohn, H. 1964: Investigations on the tropical easterly jet. *Bonner Met. Abhandl.* (4) 1–83.
Flohn, H. *et al.* 1968: Air-mass dynamics or subsidence processes in the Arabian Sea summer monsoon. *Jour. Atmos. Sci.* **25**(3) 527–529.
Fukui, E. 1942: Nanpôken no kikô (Climate of Southeast Asia). Tokyo 316p.
Ginsburg, N. S., *et al.* 1959: The pattern of Asia. Englewood Cliffs 929p.
Hann, J. 1910: Handbuch der Klimatologie. II Band, I. Teil, Klima der Tropenzone. Stuttgart 426S.
Hatakeyama, H. (ed.) 1964: Ajia no kikô (Climate of Asia). Tokyo 577p.
Johnson, B. L. C. 1969: South Asia: Selective studies of the essential geography of India, Pakistan and Ceylon. London 164p.

Kuo, C.-y., Yang, C.-s. 1965: State and structure of the atmospheric circulation in midsummer in China. *Tiri-chi-kan* (9) 69–84.

Miller, A. A. 1959: Climatology. London 318p.

Miller, F. R., Keshavamurthy, R. N. 1968: Structure of an Arabian Sea summer monsoon system. East West Center Press, Honolulu 1–94.

Murakami, T. 1959: The general circulation and water-vapour balance over the Far East during the rainy season. *Geoph. Mag.* **29** 131–171.

Ogasawara, K. 1945: Nanpô kikôron (Climates of Southeast Asia). Tokyo 396p.

Passerat, C. 1906: Les pluies de mousson en Asie. *Ann. de Geogr.* **15** 193–212.

Ramage, C. S. 1966: The summer atmospheric circulation over the Arabian Sea. *Jour. Atmos. Sci.* **23**(2) 144–150.

Ramanathan, K. R. 1960: Monsoons and the general circulation of the atmosphere. *In* Monsoons of the World 53–64.

Robinson, H. 1966: Monsoon Asia. London 559p.

Sadler, J. C. 1969: Average cloudinesss in the tropics from satellite observation. East-West Center Press, Honolulu 23p+figs.

Saito, N. 1966: A preliminary study of the summer monsoon of southern and eastern Asia. *Jour. Met. Soc. Japan II* **44** 44–59.

Sandoval, A. 1967: Background studies for a climatology of the intertropical convergence zone in the western central Pacific area. The Univ. of Wisconsin, Ph. D. Thesis, 127p.

Shen, Ch.-ch. 1965: Three-dimensional structure of the prolonged dry period in the Kiang Hwai region in July, 1959. *Tiri-chi-kan* (9) 19–32.

Sion, J. 1928: Asie des moussons. Paris 272p.

Spate, O. H. K. 1957: India and Pakistan. London 829p.

Stamp, L. D. 1962: Asia, a regional and economic geography. New York 730p.

Supan, A. 1898: Die Verteilung des Niederschlags auf der festen Erdoberfläche. Petermanns Mitt. Ergänzht 124.

Thompson, B. W. 1951: An essay on the general circulation of the atmosphere over Southeast Asia and the West Pacific. *Q. Jour. Roy. Met. Soc.* **77** 569–597.

Trewartha, G. T. 1962: The earth's problem climates. Madison 334p.

Watts, I. E. M. 1955: Equatorial weather. London 224p.

Woeikov, A. 1887: Die Klimate der Erde. 2. Teil.

Woeikov, A. 1907: A propos de l'article de Mr. Passerat sur les pluies de mousson en Asie. *Ann. de Geogr.* **16** 360–361.

Yao, Ch.-sh. 1958: The variability of precipitation in Eastern China. *Acta Met. Sinica* **29** 225–238.

Yao, Ch.-sh. 1965: Statistical methods in climatology. Peking 246p.

Yoshino, M. M. 1963: Rainfall, frontal zones and jet streams in early summer over East Asia. *Bonner Met. Abhandl.* (3) 1–127.

Yoshino, M. M. 1965, 1966: Four stages of the rainy season in early summer over East Asia (Pts. I and II). *J. Met. Soc. Japan II* **43** 231–245, **44** 209–217.

Yoshino, M. M. 1967: Atmospheric circulation over the northwest Pacific in summer. *Met. Rdsch.* **20**(2) 45–52.

Yoshino, M. M. 1969: Climatological studies on the polar frontal zones and the intertropical convergence zones over South, Southeast and East Asia. *Climatol. Notes, Hosei University, Tokyo* (1) 1–71.

PART IV

SECULAR VARIATION OF PRECIPITATION
OR CLIMATIC CHANGE OVER MONSOON ASIA

REGIONALITY OF SECULAR VARIATION IN PRECIPITATION OVER MONSOON ASIA AND ITS RELATION TO GENERAL CIRCULATION

Minoru YOSHIMURA

Abstract: Absolute values of precipitation at 92 stations in the area concerned in January and July were examined. Mean values of the three highest peaks and three lowest bottoms at each station were calculated and maps were drawn from these. Using the results, and considering the distribution of rainy or dry areas in both months, the whole area was divided into 17 regions for each month. After calculation of the correlation coefficients for precipitation, this was corrected to 17 regions for January and 20 regions for July. It was found that the boundaries of the divisions, both the original and the corrected ones, are not only similar but also correspond closely with the frontal zones or weather systems. Lastly, inter-regional relations were discussed, based on the ratio of the number of station-pairs with significant correlation to the total number of combined station-pairs in each combination of two regions. Generally speaking, the number of inter-regional combinations with a ratio under 1/10 is quite large; however, there are a considerable number of combinations with a ratio over 1/6 in winter and over 1/7 in summer. Among the latter regions, many have a significant correlation to the Siberian high or Pacific polar frontal zone in winter, but in summer, the number of regions showing such a correlation is smaller and their locations are more scattered. Apparently in summer the distribution is more complex in its relation to the atmospheric circulation than in winter.

FOREWORD

The greater part of Monsoon Asia is known as one of the rainiest regions of the world (Hatakeyama, 1964). But the map of relative variability in annual precipitation given by Koeppe and de Lang (1958) shows that this area has a greater variability than any other region with nearly the same annual precipitation. This situation seems to be produced by the alternation of relatively dry years with humid years in Monsoon Asia. Other characteristics of the precipitation and of the relative variability of annual precipitation have relatively great regional differences in this area. This complexity may be ascribed to the combined elements of varied landforms, distribution of land and sea and peculiar wind systems as explained by Garbel (1958). One of the most important means of clarifying such characteristic is to analyze the secular variation and regionality of the precipitation. From a practical viewpoint, this also contributes to an evaluation of the water resources of Monsoon Asia.

1. Earlier Studies on the Secular Variation of Precipitation and Its Regionality

a. *Annual Precipitation*

Brooks (1921) gives a map showing the trend of annual precipitation throughout the world since 1870. The map is divided into three regions, showing increasing, decreasing and intermediate precipitation. Monsoon Asia north of 35°N, with the exception of the Japanese islands, belongs to the increasing region. The Indian peninsula and the islands around

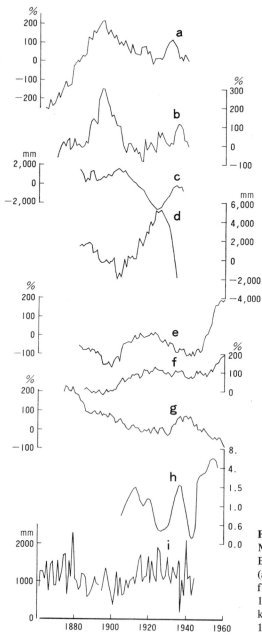

Fig. 1. Secular variation of annual rainfall in Monsoon Asia. a: at Ziazgpatan, b: at Jaipur (after E. R. Kraus, 1955), c: at Pontiak, d: at West Java (after K. J. ten Hoopen *et al.*, 1951), e: at Kanazawa, f: at Kagoshima, g: at Hakodate (after T. Sekiguti, 1963), h: mean value of east China (after Hsien-kung, Chan, 1963), and i: at Seoul (after H. Arakawa, 1956).

Indonesia tend to have decreasing precipitation. The rest of the area corresponds to the intermediate region. Since then, there have been few studies on these problems until Lamb's work in 1966.

Investigations on the regionality of rainfall variation have been made in several countries in Monsoon Asia itself. Pramanik and Jagannathan (1952) mentioned the trend of rainfall in their studies on climatic changes over the Indo–Pakistan area. According to them, among thirty stations in the area, Kozhikode (12°N, 76°E), Bellary (15°N, 77°E) and Nagpur (22°N, 79°E) showed an increasing trend, while Vishakhapatnam (18°N, 83°E) had a decreasing trend. However, a statistical test for 10-year running-means indicates that there is no trend towards either increase or decrease at these four stations. For Indonesia, ten Hoopen and Schmidt (1951), using the 11-year running-total of precipitation in Pontianak and West Java from 1899 to 1940, found a correlation ratio of −0.75 between them.

Applying the residual mass curve method, Sekiguti (1963) pointed out that the main part of Japan could be divided into three types: a) The Pacific side type, which has two peaks, one in the second decade and one towards the end of the fourth decade of this century. b) The Hokkaido type, which is very much the opposite of the Pacific type. c) The type along the coastal area of the Japan Sea, which has a variation similar to that of the Pacific side, regardless of its amplitude.

Figure 1 compares the results of ten Hoopen and Schmidt (1951), Sekiguti (1963), Kraus (1955), Arakawa (1956), and Chang et al. (1963). According to the secular variation in annual precipitation, the area can be divided into two groups, one including the stations in India and West Java and the other the stations along the Pacific side of the Japanese Islands, with Seoul in Korea and Pontianak in Borneo. The curve shown by Chang, obtained from data from many stations in a wide area of eastern China, has a trend similar to that along the Pacific side of Japan, except for the period in the fourth decade of this century. The curve for Hokkaido, Japan, seems to be the same as that for West Java, with the exception of the period before 1890.

The decreasing trend in the tropical area was initially explained by a shortening of the rainy season, which was caused by a narrowing of the ITC and the rapidity of its seasonal shifts (Kraus, 1955; ten Hoopen et al., 1951). But if this were true, the curves for Vizagpatam, Pontianak and West Java should show identical trends, judging from the positions of the ITC (Yoshimura, 1970a). A recent study by Lamb (1966), referring to the mean annual precipitation from 1931 to 1960, pointed out that the mean precipitation from July 1961 to June 1964 had a decreasing trend in an extensive area from China to East Pakistan and in West Pakistan, and an increasing trend in India, Kashmir, the Japan Sea coast, and the Korean peninsula. These results differ from the conclusions derived from Fig. 1. According to Lamb, the recent variation in precipitation was caused by a change in position of trough and ridge in the middle latitudes.

b. *Summer Precipitation*

Many studies have been made on precipitation in summer, when the greater part of Monsoon Asia lies under the influence of the rainy season. The secular variation of precipitation in summer is more complicated than that in other seasons. According to Pramanik (1952), a decreasing trend was found at Vishakhapatnam and an increasing trend at Kozhikode, Bellary, Nagpur, and Jalpargur. For India, Subbaramayya (1968) investigated a parallelism of secular variation during the season and came to the conclusion that there existed a negative correlation between West and Northeast India (Fig. 2a). For eastern China, Chang,

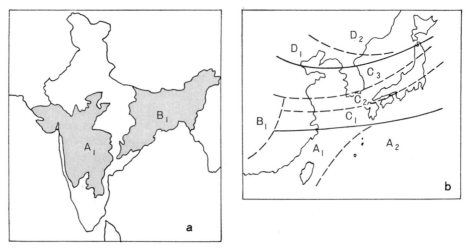

Fig. 2. Regional division based on parallelism of summer rainfall.
a. After I. Subbaramayya, 1968. b. After M. M. Yoshino, 1963.

et al. (1963) investigated the distribution pattern of signs of deviation from the normal precipitation in 1901–1960 and reported that the pattern of deviation near 33°N was the most striking of all 11 types. In a study of precipitation over the Japanese islands in June, Suda (1963) found that there were two peaks of precipitation around 1910 and 1950, and two troughs around 1890 and 1930. Comparing three decade means for 1931–1960 with the mean of the same period, he pointed out that the value of the third decade was smaller than that of the mean except in Hokkaido, and thereafter other areas of the Japanese islands maintained an increasing trend. However, Sekiguti (1963) found no apparent changes in the trend of summer precipitation over the Japanese islands in this century. Yoshino (1963), after analyzing the parallelism in secular variation in precipitation for June and July in East Asia, divided that area into four main regions with sub-regions. He noted, among other things, that the stations in Region C_1 (see Fig. 2b), extending from the lower Yangtze to western and southern Japan, showed closely parallel features with respect to precipitation in July.

c. *Winter Precipitation*

Nagpur, the only station showing an increasing trend in India, was analyzed by Pramanik (1952). The line of 33°N again seems to have significance in eastern China in the winter (Chang, 1963). For the area along the Pacific side of Japan, Sekiguti (1963) pointed out that after an increase to 1937 or 1938, a static situation persisted until the beginning of a second increase towards the end of the 1940's. During that period, however, he could find no significant features along the coastal area of the Japan Sea. In Hokkaido, except for a period with relatively small precipitation from 1900 to 1920, an increasing trend has continued since around 1950. In an analysis of the secular variation of January precipitation in the northern Pacific, Yoshimura (1970b) showed that, while the northern part of the Korean peninsula and inland part of the Asian continent had a tendency similar to that of the Pacific side of Japan and the coastal area of China, the northern part of the coastal area of the Japan Sea showed a reverse trend. He stressed the importance of the movement of frontal zones to make clear the regionality of secular variation in precipitation.

Reviewing the preceding studies, the following difficulties remain: 1)   Lack of information on precipitation around the Indochina peninsula makes it impossible to trace the regional relations covering the whole area of Monsoon Asia.   2)   Due to the fact that the secular change of annual precipitation does not keep pace with the change in precipitation during the rainy season, it is difficult to assess regional differences in climatic changes.   3) A scarcity of studies on regional differences in precipitation here makes it difficult to obtain useful information.

2. Materials and Method of Analysis

The area treated in the present paper comprises the Asian continent east of 67°E, west of 146°E, north of 6°S and south of 45°N except for the area north of Himalaya and the northwestern part of China. The data for the stations in the area shown in the appendix were taken Data Sources (1)–(5).

a. *Regional Division by Max. 3 and Min. 3*

The amplitude of secular variation at a station for the months of January and July was obtained by the following procedure: 1)   The mean value of the first three from the maximum in each month was calculated and termed "Max. 3."   2) The mean value of the last three from the minimum in each month was calculated and termed "Min. 3." The distribution of Max. 3 and Min. 3 in January and July is shown in Figs. 3a and 3b and Figs. 4a and 4b respectively. Taking the distribution pattern as a basis, the whole area is divided into 17 regions for the two months.

1)   In January, the area and situations to be noted in each region are as follows:

Region 1.   Inland continental Asia. Max. 3 is less than 50 mm and Min. 3 is 0 mm.

Region 2.   Northeastern continental Asia. Max. 3 is less than 50 mm and Min. 3 is 0 mm. The stations in this area are similar in variation trend of precipitation to the stations along the Pacific coast of Japan (Yoshimura, 1970b).

Region 3.   The southern side of Himalaya. Max. 3 is more than 100 mm and Min. 3 is 0 mm.

Region 4.   The western part of India and West Pakistan. Max. 3 is less than 50 mm and Min. 3 is 0 mm.

Region 5.   The central part of the Indian peninsula. Max. 3 is from 30 to 100 mm and Min. 3 is 0 mm.

Region 6.   The coastal area of the Bay of Bengal. Max. 3 is from 20 to 70 mm and Min. 3 is 0 mm.

Region 7.   The area centering around the mountainous region of the Indochina peninsula. Max. 3 is more than 50 mm and Min. 3 is 0 mm.

Region 8.   The main part of China and the Korean peninsula. Max. 3 is from 50 to 100 mm, in part exceeding 200 mm. Min. 3 is more than 0 mm.

Region 9.   Japanese coastal area along the Japan Sea. Max. 3 is from 150 to 600 mm and Min. 3 is from 25 to 100 mm.

Region 10.   The southern half of the Indochina peninsula Max. 3 is less than 100 mm and Min. 3 is 0 mm.

Region 11.   The Pacific islands around the Philippine islands. Max. 3 is more than 100 mm and Min. 3 is about 50 mm.

Region 12.   The Pacific coast of Japan and the adjacent area in the Pacific. Max. 3 is more than 200 mm and Min. 3 is more than 25 mm.

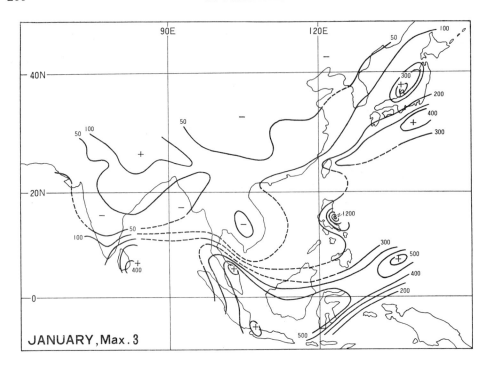

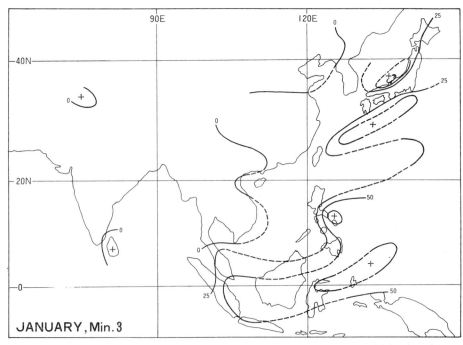

Fig. 3. Areal distribution of Max. 3 and Min. 3 in January and regional division based on these values.

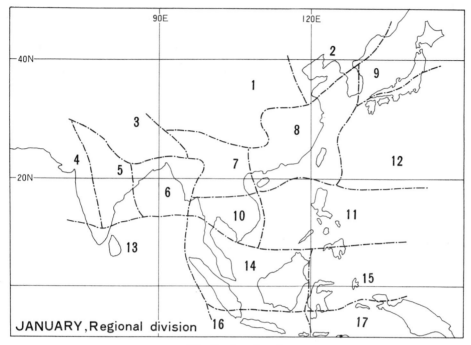

Fig. 3.   (continued)

Region 13.   The southern part of the Indian peninsula and Ceylon. Max. 3 is more than 50 mm and Min. 3 is about 20 mm.

Region 14.   The islands around the Java Sea. Max. 3 is more than 100 mm and Min. 3 is from 0 to 25 mm.

Region 15.   East of the Makassar Strait. Max. 3 is from 200 to 500 mm and Min. 3 is about 150 mm.

Region 16.   The Javanese islands. Max. 3 is 500 mm and Min. 3 is less than 50 mm.

Region 17.   The area around the Banda Sea. Max. 3 is less than 200 mm and Min. 3 is less than 50 mm.

2)   In July, the area and situations to be noted in each region are as follows:

Region 1.   The area centering around Manchuria. Max. 3 is more than 200 mm and Min. 3 is less than 25 mm.

Region 2.   East of Region 1, including the coastal areas of the continent and most of Hokkaido. Max. 3 is less than 200 mm and Min. 3 is from 25 to 50 mm.

Region 3.   The greater part of dry inland Asia. Max. 3 is less than 200 mm and Min. 3 is 25 mm.

Region 4.   Eastern China north of the Yangtze. Max. 3 is more than 200 mm and Min. 3 is less than 25 mm. The amplitude of the secular variation in this area is remarkable.

Region 5.   The Korean peninsula and the western part of Tôhoku district, Japan. Max. 3 is more than 400 mm and Min. 3 is more than 25 mm. This area receives among the heaviest rains in Monsoon Asia for this month.

Region 6.   The eastern part of the Japanese islands. Max. 3 is less than 400 mm and Min. 3 is less than 25 mm.

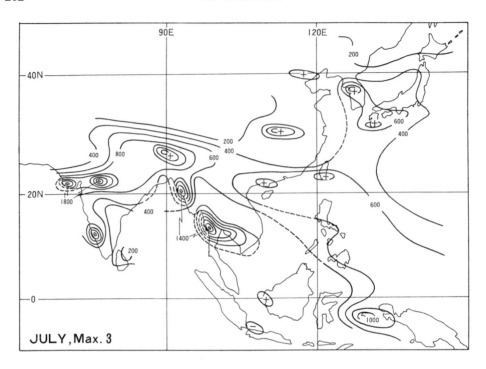

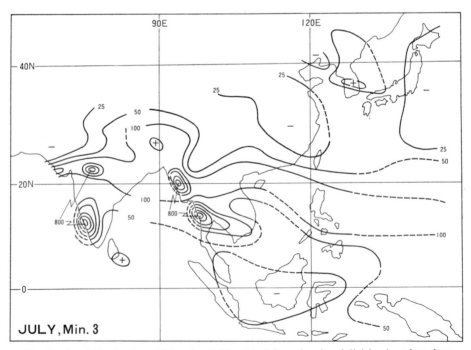

Fig. 4. Areal distribution of Max. 3 and Min. 3 in July and regional division based on these values.

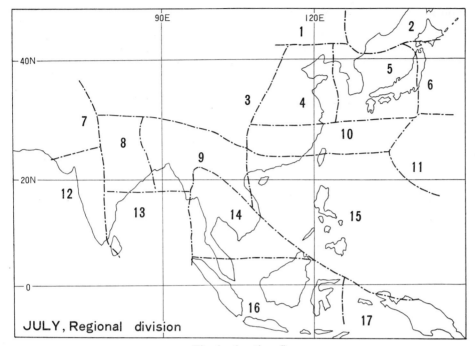

Fig. 4. (continued)

Region 7. The northern part of India and West Pakistan. Max. 3 is from 200 to 400 mm and Min. 3 is less than 25 mm.

Region 8. The northern part of Deccan. Max. 3 is more than 600 mm and Min. 3 is from 50 to 100 mm. This area receives less precipitation in the summer monsoon than its neighbors to the east and west.

Region 9. The northeastern part of India, East Pakistan and the northern part of the Indochina peninsula. Max. 3 is more than 400 mm, partly exceeding 1,200 mm. Min. 3 is less than 100 mm.

Region 10. The southern part of China and the Ryukyu islands. Max. 3 is from 300 to 500 mm and Min. 3 is from 25 to 50 mm.

Region 11. The Pacific islands north of 20°N. Max. 3 is less than 400 mm and Min. 3 is from 25 to 50 mm.

Region 12. The western part of Deccan and the Laccadive islands. Max. 3 is 400 mm, partly exceeding 1,300 mm. Min. 3 is more than 50 mm, partly exceeding 800 mm.

Region 13. The eastern part of Deccan and Ceylon. Max. 3 is less than 400 mm and Min. 3 is less than 50 mm.

Region 14. The greater part of the Indochina peninsula. Max. 3 is more than 1,400 mm and Min. 3 varies according to the location.

Region 15. The islands in the Yellow Sea, the Philippine and Palau islands in the Pacific. Max. 3 is more than 600 mm and Min. 3 is from 50 to 100 mm. This region is seriously affected by typhoons.

Region 16. The islands around the Java Sea. Max. 3 is less than 400 mm and Min. 3 in near 50 mm.

Region 17.   The Ceram islands and vicinity. Max. 3 is more than 1,000 mm and Min. 3 is more than 50 mm.

It must be added that the trend of secular variation is not always uniform within each region, because the division was made according to the amplitude of precipitation variation.

When we compare the distribution of Max. 3 and Min. 3 in July with that of "high" and "low" in the same month by Yohsino (1963), there is some interesting coincidence between the distribution patterns of Max. 3 and of "high", despite the presence of some isopleth of lesser degree, including three rainy areas in East Asia. There is a general coincidence between the distribution patterns of Min. 3 and of "low"; however, we should note the appearence of "low" areas in inland China and from the southern Pacific side of the Japanese islands to the Ryukyus.

b.   *Parallelism of Secular Variation in Precipitation*

In analyzing parallelism in secular variation, correlation coefficient or correlation index methods are considered suitable. However, because of the difference in the time length of the available data or of temporary breaks in observations, the correlation index method was adopted to estimate the correlation coefficient.

Plus, minus or zero signs, taking the 5-year running mean for each year as base, were entered on the series of precipitation data for every station. This made it easier to avoid any difficulty in counting the mean value for different time lengths separately for combinations of two stations. Using combinations of the sign series thus obtained, the correlation index r was calculated from the following expression:

$$r = (R_1 + R_0/2)/N \tag{1}$$

R_1 is the number of years with the same sign at a pair of stations. R_0 is the number of years with a zero sign, and N is the total number of years included.

In this study, station-pairs with a correlation index greater than 0.67 are defined as having positive correlation. The exception is the case when N is smaller than 10. If the correlation index is over 0.67, it can be estimated that the correlation coefficient is larger than 0.5. When the number of years with opposite signs $(+, -)$ in a pair is greater than that of the number of years showing agreement $(+, +; -, -)$, it can be estimated by simple computation that the correlation coefficient is negative. Correlation indices for all 4,140 pairs of stations were then calculated for the months of January and July.

The area distribution of combined station-pairs having significant correlation is so complicated that it is impossible to divide the area into a simple pattern of regions. Therefore, a correction was applied to make a new division, following three conditions: 1)   One or both stations of the station-pair with negative correlation in a region are moved from the original region to another, according to the situation in positive correlation between each of the pair and other stations within each original station.   2)   When there is a positive correlation between two stations in neighboring regions, one of the two stations is moved from its original region to another, according to the number of positive correlations of each station to other stations in its own region.   3)   When there is no correlation between two stations in neighboring regions, and both stations have a correlation, either positive or negative, to a station in an other separate region, both stations remain in their original region or are removed to the neighboring region, according to the correlation between the other stations in each of two regions to the station in a separate region.

Results for the two months are shown in Figs. 5 and 6 respectively.

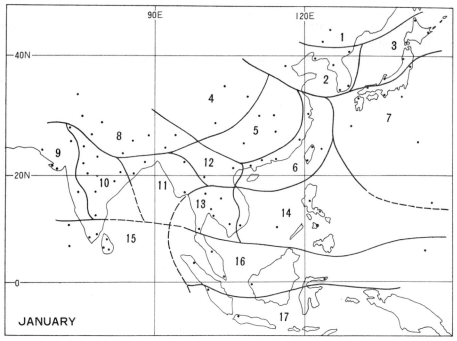

Fig. 5. Areal division based on parallelism of secular variation of January precipitation.

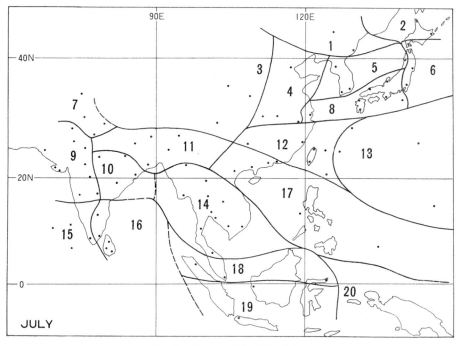

Fig. 6. Same as Fig. 5 for July.

c. *New Division*

1)   Division of the area into 17 regions for January, as shown in Fig. 5, differs from the former division shown in Fig. 3c.

Region 1.   NE Monsoon Asia centered around Manchuria.
Region 2.   The Korean peninsula and the northern part of China.
Region 3.   The coastal area along the Japan Sea and Hokkaido.
Region 4.   Inland continental Asia.
Region 5.   The inland part of China.
Region 6.   Coastal China along the South China Sea and the Yellow Sea.
Region 7.   The Pacific side of Japan with its adjacent islands.
Region 8.   The southern side of Himalaya.
Region 9.   The western part of Deccan and West Pakistan.
Region 10.   The eastern part of Deccan.
Region 11.   The eastern part of India, East Pakistan and Burma.
Region 12.   The northern part of Laos and North Vietnam.
Region 13.   The southern part of the Indochina peninsula except the coastal area along the South China Sea.
Region 14.   The Philippines and their adjacent islands in the Pacific.
Region 15.   The southern part of the Indian subcontinent and Ceylon.
Region 16.   The zonal area north of the equator.
Region 17.   The area south of the equator.

2)   Division of the area into 20 regions for July, which is shown in Fig. 6, is more complicated than that of January.

Region 1.   Manchuria and the coastal area along the northern Japan Sea.
Region 2.   The greater part of Hokkaido, Japan.
Region 3.   The western part of China.
Region 4.   The lowlands of the Hopei district, China.
Region 5.   The Korean peninsula and the coastal area of the Tohoku district, Japan, along the Japan Sea.
Region 6.   The eastern part of Honshu, Japan.
Region 7.   Kashmir and the northwestern part of India.
Region 8.   The southern part of Japan and the Yangtze delta region.
Region 9.   The western part of Deccan and West Pakistan.
Region 10.   The eastern part of the Indian peninsula.
Region 11.   The Hindustan Plain, Assam and the Ganges delta.
Region 12.   Formosa and its adjacent islands.
Region 13.   The Bonin islands and its adjacent area in the Pacific.
Region 14.   The Indochina peninsula.
Region 15.   The southwestern part of the Indian subcontinent.
Region 16.   The southeastern part of the Indian subcontinent including Ceylon.
Region 17.   The coastal area along the Gulf of Tongking and the Philippines with their adjacent islands.
Region 18.   Borneo with its adjacent islands.
Region 19.   The islands around the Java Sea.
Region 20.   The islands in the eastern part of the Banda Sea. As there is only one station here, it is necessary to establish whether this area may be regarded as an independent region or not.

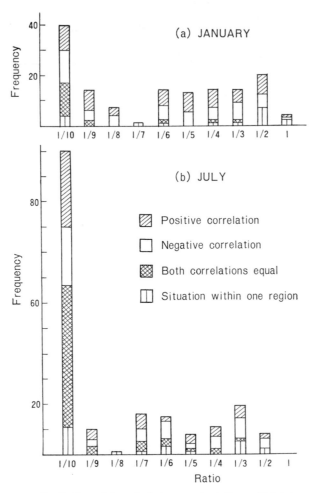

Fig. 7. Frequency distribution of the ratio between pair-stations which have more than 0.5 or less than −0.5 and the total number of station-pairs in each combination of regions.

Comparing the new divisions (Figs. 5 and 6) with the former (Figs. 3 and 4), it can be said that, despite their similarity, the amplitude of the secular variation in precipitation is reflected only in the new ones. The ratio of the number of station-pairs with correlation coefficients greater than 0.5 or smaller than −0.5 to the total number of station-pairs was counted for every combination of regions. The frequency distributions of these ratios are shown in Figs. 7a and 7b. When two different kinds of station-pairs with positive or negative correlations between two regions appear, the inter-regional relation is expressed by the sign of the prevailing number of significant correlations, either positive or negative. It will be seen that all inter-regional combinations with a positive or negative correlation, shown grouped on the right in the figures, have a tendency to change in the same or in opposite ways. The lowest limit of the right-hand group for January was determined to be 1/6 and for July 1/7, When the station-pairs with a region account for over 1/4, they are taken as the standard for that region.

In January, except for 4 regions, 3, 4, 12, and 17, which have no station-pairs with positive

Table I. Number of station-pairs having a correlation greater than 0.5 (upper) or smaller than −0.5. (lower). ++, (−−) mean the ratio is equal or greater than 1/4 and +, (−) mean the ratio is between 1/5 and 1/6.

Inter-regional relations (Jan.)

Region	1	2	3	4	5	6	7	8	9	10	11	12	13	14	15	16	17
1	++ 1 0																
2	++ 3 2	++ 10 0															
3	1 1	3 0	0 0														
4	0 0	+ 3 0	1 0	0 0													
5	++ 3 1	− 1 5	1 0	0 1	++ 11 0												
6	0 2	5 2	0 2	0 5	− 1 3	++ 11 0											
7	+ 3 0	7 0	0 1	1 0	2 2	+ 12 1	++ 8 0										
8	2 0	+ 11 2	0 0	++ 8 0	2 0	3 0	1 1	++ 10 0									
9	++ 4 0	+ 5 1	1 1	++ 4 1	++ 5 1	5 3	4 1	++ 18 0	++ 4 0								
10	++ 7 0	4 3	0 3	+ 5 0	1 1	5 4	0 4	+ 16 1	++ 22 0	++ 14 0							
11	++ 3 0	2 1	1 0	1 0	− 1 3	1 2	1 0	+ 6 1	++ 8 1	6 0	++ 3 0						
12	−− 0 1	− 0 0	−− 1 2	1 2	−− 1 3	1 2	1 3	2 2	3 3	+ 4 3	++ 3 1	0 0					
13	− 0 2	1 1	− 1 5	3 3	3 2	6 1	2 3	6 0	+ 5 3	5 4	+ 5 1	−− 2 4	++ 7 0				
14	0 1	2 2	0 0	−− 0 3	1 2	2 2	4 1	3 2	2 1	2 1	+ 1 0	+ 2 1	+ 6 1	1 0			
15	− 0 4	4 1	0 1	4 1	0 4	4 3	1 4	0 3	5 5	6 3	−− 6 1	+ 1 5	8 1	0 4	+ 10 0		
16	− 0 2	4 4	2 0	1 0	−− 1 8	0 3	1 3	1 2	+ 5 4	3 4	+ 5 1	0 3	2 4	0 3	+ 8 2	+ 2 0	
17	++ 1 0	−− 1 6	0 0	−− 0 3	−− 0 3	1 0	0 0	− 0 3	−− 1 3	−− 1 4	0 0	1 1	− 2 1	0 2	− 2 4	− 2 1	0 0

correlation, more than 1/6 of the pairs have positive correlation, and more than 1/4 of all the pairs in 10 regions, except for Regions 14, 15, and 16, have positive correlation. Various situations between regions in January are shown in Table I.

Inter-regional combinations with positive correlation for station-pairs are between Regions 1–5 and between Regions 7–11; the relation between Regions 8 and 11 is a typical case. While station-pairs in Region 12 have negatve correlation to the pairs in almost all other Regions, the pairs in Region 14 have positive correlation only to those in neighboring regions. The characteristics of Region 17 should be noted, as also the fact that the pairs in Region 2–5 have a negative correlation to those in Regions 13–16.

In July, there are fewer regions which have station-pairs with significant correlation. More than 1/4 of all pairs in Regions 1, 5, 8, 10, 13, 16, and 19, and more than 1/7 but less than 1/4 of all pairs in Regions 4, 9, 12, and have positive correlation. Inter-regional relations in July are given in Table II.

Table II. Number of station pairs having a correlation greater than 0.5 (upper) or smaller than −0.5 (lower). ++, (— —) mean the ratio is equal or greater than 1/4 and +, (−) mean the ratio is between 1/5 and 1/7.

Inter-regional relations (July)

Region	1	2	3	4	5	6	7	8	9	10	11	12	13	14	15	16	17	18	19	20
	++																			
1	1																			
	0																				
	— —																				
2	0	0																		
	1	0																			
	+	−																			
3	1	0	0																	
	0	1	0																		
				+																
4	0	0	1	1																
	1	0	0	0																
		−		— —	++															
5	0	0	0	0	1															
	0	1	0	4	0															
		−	−		++															
6	0	0	1	1	3	0														
	0	1	2	1	0	0														
			−																	
7	0	0	1	1	0	0	0													
	1	0	2	0	0	0	0													
	−	−	−		+			++												
8	1	1	0	1	3	1	0	6												
	3	3	4	1	0	0	0	0												
		— —	++							+										
9	1	0	9	2	1	1	2	0	10											
	1	3	1	0	2	0	0	3	0											
				+						++										
10	0	0	2	3	0	0	0	0	1	7										
	0	0	2	0	0	0	0	0	2	0										
	−			+																
11	0	1	1	3	1	0	0	0	2	0	0									
	2	0	1	1	1	0	0	0	2	0	0									
	— —												+							
12	4	1	1	1	2	1	0	1	5	0	1	3								
	5	0	1	0	0	0	1	0	6	1	2	0								
	— —				+	+							++							
13	1	1	0	1	2	2	0	2	1	0	1	3	2							
	3	1	0	1	0	0	0	0	0	0	0	1	0							
14	0	0	1	1	0	0	0	0	4	0	0	1	0	2						
	2	0	1	0	0	0	0	0	0	0	0	2	1	0						
				+	−								+		−					
15	1	1	2	4	2	1	0	2	1	1	2	2	3	2	1					
	1	0	2	3	3	0	0	1	2	0	1	2	1	0	0					
			++	++								−	++			++				
16	1	0	6	8	1	0	2	2	6	2	0	0	5	1	2	5				
	0	1	2	4	2	0	1	1	1	0	1	7	1	0	2	0				
	++			−		+	+									++				
17	3	1	2	2	2	4	4	1	2	1	1	1	0	2	3	2	2			
	1	1	1	5	0	0	0	2	3	1	5	1	1	0	0	10	0			
	— —		— —		++			++									−			
18	0	0	0	0	3	0	2	6	1	0	1	0	0	0	1	1	0	0		
	1	3	0	1	0	1	2	0	0	1	0	0	0	1	1	0	3	0		
	— —		— —		— —						++				+	— —	++		++	
19	0	0	0	0	1	0	0	2	1	0	4	2	0	0	3	0	3	0	1	
	3	0	3	1	2	0	1	0	2	1	0	2	0	2	1	6	1	0	0	
	++		— —		−			−										++		
20	1	0	1	1	0	0	0	0	0	0	0	1	0	0	0	0	1	2	0	0
	0	1	1	1	2	0	0	1	0	0	0	1	0	0	1	1	1	1	0	0

The tendency mentioned by Subbaramayya (1968) does not appear between Regions 9 and 10, or 11. Regions 9 and 15 are sometimes treated as one region; however, the latter has a positive correlation to Regions 4, 13, and 19.

Figs. 8 and 9 illustrate the secular change of precipitation in January and in July, taking one station out of each region. The tendencies mentioned above are also found in the long range precipitation regime. For example, among Regions 1, 2, and 5, which form a group having a number of station-pairs with positive correlation, there is a similarity in the regime of station-pairs with positive correlation, there is a similarity in the regime of precipitation

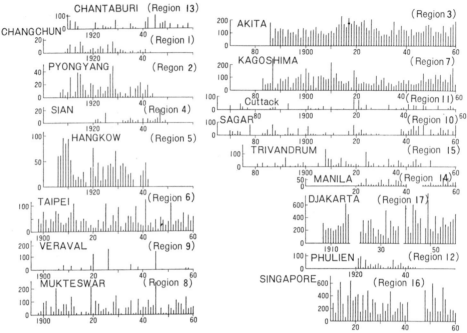

Fig. 8. Secular variation of rainfall in January in each region.

for January; all three stations in the three regions have peaks in the first half of the 1910's, in the second half of the 1920's and around 1940, and troughs toward the end of the 1910's and around 1935. Other features can be noted in both figures.

d. *Some Comments on the Meaning of Regional Division*

In order to consider the regional division mentioned above, the results are discussed in relation to those obtained by Yoshino (1969) and Yoshimura (1970a, 1970b) from the view point of general circulation.

In January, Region 1 is always under the control of the Siberian high. But in Region 2, the Eurasian polar frontal zone sometimes appears in the southern part of Japan in warm winters (Yoshimura, 1970b). The area south of this frontal zone to north of the Pacific polar frontal zone is also controlled by the Siberian high, when it develops. Therefore, it is quite natural that a positive correlation should exist between Regions 1 and 5. The existence of negative correlation between the stations in Region 5 and those in Regions 2 and between the stations in Region 6 and those in Regions 4 can be explained by the development of frontal zones in Regions 2, 6, and 7 in warm winters. The situation is also explained by the existence of positive correlation between the stations in Regions 6 and 7. Positive correlation between Regions 8, 9, 10, and 11 and regions mentioned above and between Regions 8, 9, 10, and 11 are explained in part by the dryness of these four regions as shown in Fig. 5. According to equation (1) given above, the occurrence of a long series of rainless years makes the correlation index r larger. Though the dryness in Regions 8, 9, 10, and 11 seems to have certain relations to the development of a Western Disturbance, no positive correlations can be found between the Siberian high and Western Disturbance, at least not in our study of the correlation between the mean pressure in northern India and in inland Siberia.

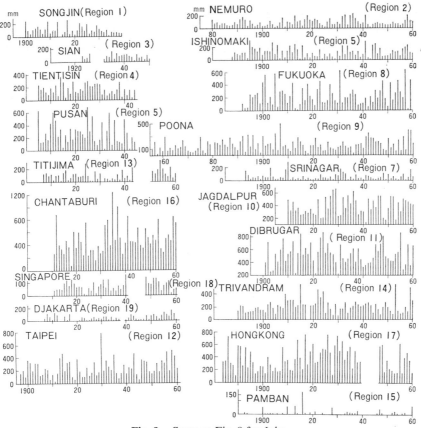

Fig. 9. Same as Fig. 8 for July.

Negative correlation between Regions 14, 15, and 16 and regions north of the Pacific polar frontal zone is ascribed to the influence of prevailing winds in winter. As trade winds blowing NE prevail in the former regions and polar winds blowing NW are dominant in the latter, winds at the stations in the former regions appear to blow landward from the sea. Apparently intrusion of the NW monsoon into the area south of the polar frontal zone, crossing the zone and shifting its direction from northwest to northeast, strengthens the NE monsoon and trade winds, and greatly increases the precipitation in Regions 14, 15, and 16.

We might add here that the boundary between Regions 16 and 17 in Fig. 5 seems to have a close relation to the ITC in January, as reported by Yoshino (1969); the boundary line between the two regions runs along the center line of the ITC.

In July, the Eurasian polar frontal zone tends to develop near the southern marginal line of Regions 1 and 2. This fact explains why the stations in both regions tend to have a negative correlation to the stations in the southern regions. The reason for the remarkable positive correlation between the station-pairs in Region 3 and Region 9, and the similar relation between Region 4 and Regions 10, 11, 15, and 16 remains unclear.

Subbaramayya (1968) states that when the SW monsoon prevails over all of India, the rainy area covers the whole subcontinent. However, when the monsoon weakens, the rainy area is reduced to a limited region in the eastern part. If the depth of the monsoon trough is reflected by the pressure at Srinagar, te pressure must show a positive correlation to precipitation at the stations in Region 4 and a negative correlation to precipitation at the stations in Region 3, judging from the relations mentioned above. We find a positive correlation greater than 0.5 between the mean pressure at Srinagar and the precipitation at Lüta, and a negative correlation less than −0.5 between the mean pressure at Srinagar and the precipitation at Sian in China. It is interesting that there are two pairs with positive correlation between India and northern China, which might be caused by the northwestward expansion of areas with a large amount of precipitable water in India and also by the north or northwestward expansion in northern China from June to July (Kubota, 1969).

The positive correlation between the stations in Region 13 and those in Regions 5, 6, 12, and 17 can be understood partly from the features around the northwestern margin of the North Pacific high. For example, correlation coefficients less than −0.5 are found between the mean pressure at Hachijo island and the precipitation at Choshi and between the mean pressure at Hachijo and the precipitation at Tainan. Referring to the stream lines shown by Yoshino (1969), positive correlation at the station-pairs between Regions 5, 8, and 18 can be easily explained by the transport of precipitable water in July. Why there are no regions with significant correlation to Regions 5 or 8 and Region 18 is an interesting problem which might be solved from the viewpoint of synoptic climatology.

The negative correlation between Regions 19 or 20 and other regions may also be related to the development of SITC in the area east of Celebes (Yoshino, 1969).

3. SUMMARY

The parallelism of secular variation in precipitation over Monsoon Asia in January and July was analyzed and the following points were made:

1)   Monsoon Asia was divided into 17 regions as shown in Figs. 3c and 4c by the amplitude of secular variation in precipitation.

2)   The above division was improved by the distribution of correlation coefficients which were estimated from the correlation index. As a result, Monsoon Asia was divided into 17 regions in January and 20 regions in July, with slight differences from the former division.

3)   Regionality was discussed after calculation of the ratio of the number of station-pairs with significant correlation to the total number of pairs. Ratios greater than 1/6 in January and 1/7 in July were regarded as the lower limit in estimating parallelism or a reverse tendency of precipitation for these regions. A ratio greater than 1/4 was taken giving a remarkable meaning to the regions. The results were presented in the tables.

4)   Roughly speaking, the positions of frontal zones, defined by the frequency distribution of actual fronts on the daily weather maps, were closely related to the area correlation of secular variation in precipitation in January. In July, the area, characteristics of secular variation were not so significant. In addition, the positions of the frontal zone seem to have little significance in explaining the regionality of precipitation in July.

References

Arakawa, H. 1956: Two minima of the average rainfall at the turn of the century and at about 1940 in the Far East. *Arch. Met. Geophys. Biokl.* [B] **7** 406–412.
Brooks, C. E. P. 1921: The secular variation of climate. *Geog. Review* **11** 120–135.

Chang, Hsien-kung, Kong, Y., Hsu, C. 1963: A preliminary analysis of abnormal precipitation distribution in eastern China in the first six decades of the present century. *Acta Met. Sinica* **33** 64–77.

Garbell, M. A. 1958: Tropical and Equatorial Meteorology. New York 237p.

Hatakeyama, H. (ed.) 1964: Climate of Asia, Kokin Shoin Tokyo 577p.

ten Hoopen, K. J., Schmidt, F. H. 1951: Recent climatic change in Indonesia. *Nature* **168** 428–429.

Koeppe, C. F., de Lang, G. G. 1958: Weather and Climate. New York-Toronto-London. 341p.

Kraus, E. R. 1955: Secular changes of tropical rainfall regimes. *Q. J. Roy. Met. Soc.* **81** 198–210.

Kubota, I. 1968: Distribution of monthly mean precipitable water in the Northern Hemisphere and its time change. *Climatological Note, Hosei Univ.* (2) 1–28.

Lamb, H. H. 1966: Climate in the 1960's: World wind circulation reflected in prevailing temperatures, rainfall patterns and the levels of African lakes. *Geogr. J.* **132** 183–212.

Pramanik, S. K., Jagannathan, P. 1952: Climatic changes in India—(1) Rainfall. *Indian J. Met. Geophy.* **4** 291–309.

Sekiguti, T. 1963: Secular change patterns of Japanese rainfall. *Geogr. Rev. Japan* **37** 217–225.

Subbaramayya, I. 1968: The inter-relations of monsoon rainfall in different subdivisions of India. *J. Met. Soc. Japan* **42** 77–85.

Suda, K. 1963: Secular change of rainfall in the Baiu season. *Tenki* **10** 119–123.

Yoshimura, M. 1970a: A climatological study on the ITCZ. *Tenki* **17** 31–38.

Yoshimura, M. 1970b: Atmospheric circulation and climatic fluctuations over the North Pacific in winter. *Geogr. Rev. Japan* **43** 139–147.

Yoshino, M. M. 1963: Rainfall, frontal zones and jet stream in early summer over East Asia. *Bonner Met. Abhandlungen* **3** 1–127.

Yoshino, M. M. 1969: Climatological studies on the polar frontal zone and intertropical convergence zone over South, Southeast and East Asia. *Climatological Note, Hosei Univ.* (1) 1–71.

Data Sources

1   Royal Obs. Hong Kong: Hong Kong Meteorological Records. 1884–1939, 1947–1962.
2   Indonesia Meteorological and Geophysical Service: Rainfall Observation in Indonesia.
3   Central Meteorological Observatory Japan 1954: Climatic Records of Japan and the Far East Area.
4   Thailand Meteorological Department: Monthly and Annual Rainfall of Thailand with Departure from Normal (for the period from 1911 to 1960).
5   U. S. Weather Bureau: World Weather Record. (1920: 1921–1930: 1931–1940: 1941–1950: 1951–1960).

Appendix: List of stations

Srinagar	34°05′N,	74°50′E	1587m	1893–1960
Karachi	24°51′N,	67°04′E	4m	1940–1960
Agra	27°10′N,	78°02′E	169m	1862–1960
Akola	20°42′N,	77°02′E	282m	1870–1960
Bikanel	28°00′N,	73°18′E	224m	1878–1960
Bangalore	12°58′N,	77°37′E	1010m	1921–1960
Cuttack	20°48′N,	85°56′E	27m	1867–1960
Calcutta	22°32′N,	88°20′E	6m	1921–1960
Daltonganj	24°03′N,	84°04′E	221m	1893–1960
Darbanga	26°10′N,	85°54′E	49m	1875–1960
Darjeeling	27°03′N,	88°16′E	2128m	1867–1960
Dibrugarh	27°28′N,	94°55′E	106m	1901–1960
Dwarka	22°22′N,	69°05′E	11m	1901–1960
Hyderabard	17°27′N,	78°28′E	545m	1893–1960
Indor	22°43′N,	75°48′E	567m	1877–1960
Jagdalpur	19°05′N,	82°02′E	553m	1909–1960
Jodhpur	26°18′N,	73°01′E	224m	1897–1960

Kota	25°11′N, 75°51′E	257m	1898–1960	
Ludhiana	30°56′N, 75°52′E	247m	1868–1960	
Mukteswar	29°28′N, 79°39′E	2311m	1897–1960	
Pamban	09°16′N, 79°18′E	11m	1893–1960	
Poona	18°32′N, 73°51′E	559m	1856–1960	
Sagar	23°51′N, 78°45′E	551m	1870–1960	
Silchar	24°49′N, 92°48′E	29m	1869–1960	
Trivandram	08°29′N, 76°57′E	64m	1853–1960	(1890)
Veraval	20°54′N, 70°22′E	8m	1893–1960	
Aminidivi	11°06′N, 72°45′E	4m	1889–1950	
Minicoy	08°06′N, 73°00′E	2m	1891–1950	
Colombo	06°54′N, 79°52′E	7m	1921–1960	
Nuwara Eliya	06°58′N, 80°47′E	1895m	1921–1960	
Trincomale	08°35′N, 81°15′E	7m	1921–1960	
Mandalay	21°59′N, 96°06′E	76m	1921–1960	(41–50)
Rangoon	16°46′N, 96°10′E	23m	1921–1960	(41–50)
Bangkok	13°44′N, 100°30′E	12m	1911–1960	
Chantaburi	12°37′N, 102°07′E	5m	1911–1960	
Chiangrai	19°55′N, 99°15′E	3m	1911–1960	
Chumphon	10°27′N, 99°15′E	3m	1911–1960	
Loei	17°32′N, 101°30′E	252m	1911–1960	
Narathiwat	06°26′N, 101°56′E	4m	1911–1960	
Phnompen	11°33′N, 104°51′E	11m	1911–1960	(41–50)
Roi Et	16°03′N, 103°41′E	1m	1911–1960	
Saigon	10°49′N, 106°38′E	106m	1921–1960	(45)
Phu-Lien	28°48′N, 106°38′E	116m	1921–1944	
Amoy	24°26′N, 118°04′E	5m	1912–1938	
Changchun	43°52′N, 125°20′E	215m	1909–1932	
Hankow	30°35′N, 114°17′E	36m	1905–1942	(37, 38)
Hongkong	22°18′N, 114°10′E	33m	1884–1960	(40–46)
I-chang	30°43′N, 111°13′E	57m	1882–1937	
Kuei-Yang	26°35′N, 106°43′E	1057m	1921–1950	
Lanchow	36°03′N, 103°56′E	1506m	1932–1950	
Lüta (Talien)	38°55′N, 121°38′E	97m	1905–1944	
Pei-hai	21°28′N, 109°05′E	4m	1885–1938	
Shanghai	31°17′N, 121°28′E	3m	1906–1950	
Mukden	41°47′N, 123°24′E	43m	1905–1942	
Sian	34°15′N, 108°55′E	395m	1923–1950	(27–31)
Tientsin	39°09′N, 117°09′E	3m	1905–1944	
Wu-Chow	23°38′N, 111°17′E	11m	1900–1946	(45)
Ya-an	30°00′N, 103°03′E	650m	1937–1950	
Songjin	40°40′N, 129°12′E	31m	1905–1944	
Pyongyang	39°01′N, 125°49′E	27m	1907–1944	
Mokpo	34°47′N, 126°23′E	31m	1904–1960	(45–50)
Pusan	35°06′N, 129°03′E	69m	1944–1944	
Ishigakijima	24°20′N, 124°10′E	6m	1897–1960	(50)
Tainan	23°00′N, 120°13′E	13m	1897–1960	
Taipei	25°02′N, 121°31′E	8m	1897–1960	
Akita	39°43′N, 140°06′E	9m	1886–1960	
Choshi	35°43′N, 140°51′E	27m	1887–1960	
Fukuoka	33°35′N, 130°23′E	3m	1891–1960	
Haboro	44°22′N, 141°42′E	8m	1921–1960	
Hachijojima	33°06′N, 139°47′E	80m	1907–1960	
Hakodate	41°49′N, 140°45′E	33m	1875–1960	
Hamada	34°54′N, 132°04′E	18m	1891–1960	
Ishinomaki	38°26′N, 141°18′E	43m	1888–1960	

Kagoshima	31°34′N, 130°33′E	4m	1883–1960	
Kanazawa	36°33′N, 136°39′E	27m	1886–1960	
Naze	28°23′N, 129°30′E	3m	1897–1960	
Nemuro	43°20′N, 145°35′E	26m	1880–1960	
Shionomisaki	33°27′N, 155°46′E	73m	1913–1960	
Titijima	27°05′N, 142°11′E	3m	1907–1960	(44–50)
Saipan	15°14′N, 145°46′E	212m	1927–1942	
Palau	07°21′N, 134°29′E	29m	1924–1960	(41–46)
Iloilo	10°42′N, 122°34′E	7m	1921–1960	(41–46)
Legaspi	13°09′N, 123°45′E	6m	1921–1960	(41–46)
Manila	14°35′N, 120°59′E	14m	1921–1960	(41–45)
Ambon	03°42′S, 128°05′E	12m	1921–1960	(18–26, 38, 39, 42–47)
Djakarta	06°11′S, 106°51′E	8m	1907–1960	(18–21, 38, 39)
Medan	03°35′N, 98°41′E	20m	1921–1960	(42–45)
Menado	01°30′S, 124°51′E	86m	1927–1960	(34–39, 42–46)
Pontianak	00°01′S, 109°20′E	3m	1912–1960	(18–26, 42–46)
Padang	00°56′S, 100°22′E	3m	1911–1960	(18–26, 38, 39, 42–46)
Singapore	01°17′N, 103°51′E	3m	1911–1960	(42–47)

FLUCTUATIONS OF RAINFALL IN SOUTHEAST ASIA— EQUATORIAL PACIFIC AND LOW AND MIDDLE LATITUDE CIRCULATIONS IN THE SOUTHERN HEMISPHERE

Iwao Tsuchiya

Abstract: Rainfall variations over a long period in Southeast Asia are very complex; however, in wide scope analyses over the region from India to the equatorial Pacific, some characteristic features are recognizable.

It is well known that year-to-year variations of rainfall over the so called equatorial Pacific dry zone which extends from the coast of Peru to 180°E or west along the equator are very peculiar. Until recently, it was not clear how these phenomena resulted from the general circulation mechanisms.

As a result of investigating the world precipitation and precipitation anomaly distribution maps included in "Die Witterung in Übersee" (1955–1968), Hamburg Seewetteramt, the author recognized that there was much less rainfall in the India–Indonesia region while there was plentiful rainfall in the dry zone. These events occurred in 1957, 1958, 1965 and 1966. In 1955, 1956 and 1962 reversal types are recognizable. Data over long periods of Indian floods and droughts and rainfall data for the Ocean and Fanning Islands in the dry zone have suggested that the reverse phase rainfall variations between India and the equatorial Pacific occurred in past years.

In addition to the relationship between these peculiar rainfall fluctuations and influences of the SE trade winds, the effect of mid-latitude westerlies in the Southern Hemisphere should be recognized because anomalous rainfall distributions occurred in 1957–1958 and 1965–1966, when southern westerlies were very weak, especially in the winter month (July).

Recently Walker's southern oscillation theory for new zonal and tropospheric circulation models in tropical and sub-tropical latitudes was advanced, and named the Walker circulation by Bjerknes (1969). It is possible to state that the variations of southern westerlies play a more important role in weakening or strengthening the Walker circulation then in modifying rainfall in Southeast Asia or the Indian summer monsoon rainfalls through variations of sea surface temperatures under the SE trade winds in the South Pacific and the SE trade winds in the Indian Ocean.

1. Introduction

Rainfall variations in Southeast Asia show very complex features. At first sight it would seem that there are no definitive relationships between any regions in Southeast Asia as observed during recent years, although many papers have dealt with this question (Pramanik

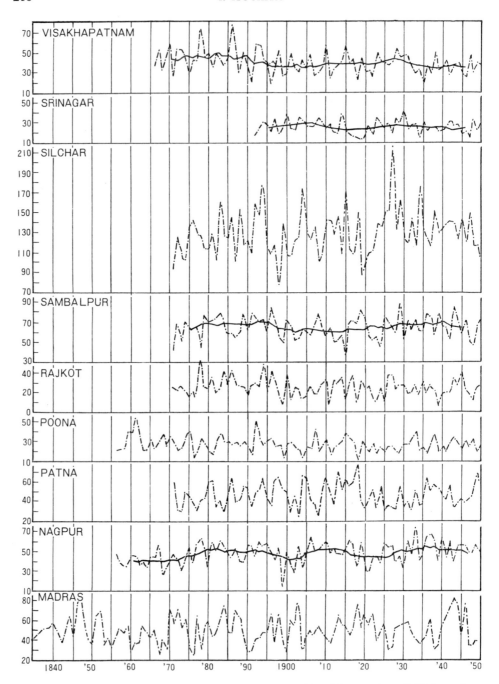

Fig. 1. Long-period rainfalls in India (based on Pramanik and Jagannathan, 1953) (annual rainfall in inches; ———10-year moving average).

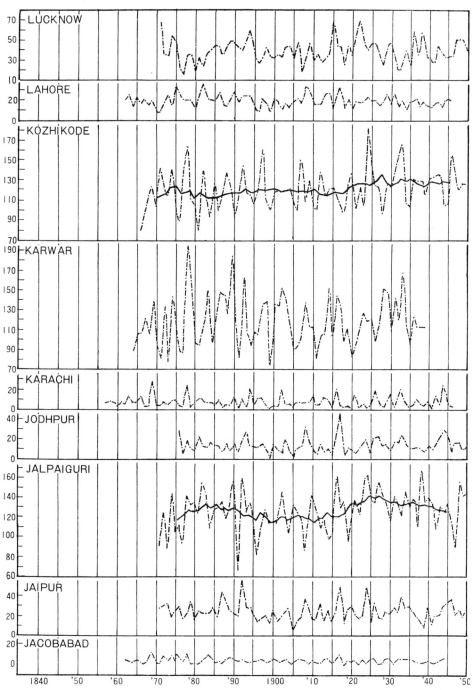

Fig. 1. (continued)

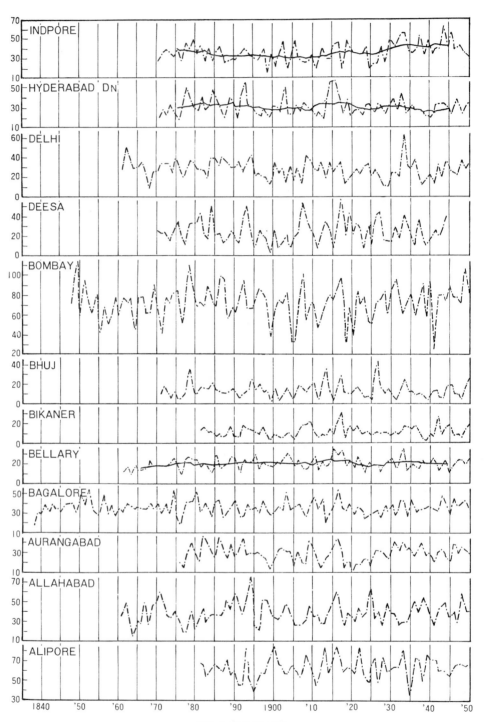

Fig. 1. (continued)

Fig. 1. (continued)

and Jagannathan, 1953; ten Hoopen and Schmidt, 1951; Sutrisno, 1964; Chen and Chen, 1965).

Because of these complexities, the forecasting method for the Indian summer monsoon rainfall, based partly on Walker's southern oscillation (Walker, 1924) has not been satisfactory so far. However, his hypothesis, which was based on some tenuous physical observations, has become a new global analysis problem since plentiful meteorological and oceanographical observational data have become available in recent years.

Troup (1965) recognized that southern oscillation is an exchange of air between the Eastern and Western Hemispheres. In his opinion, the oscillation should not be characterized by any one period, but should simply be regarded as a standing fluctuation. Berlage (1966) introduced the atmospheric pressure at Djakarta, Indonesia as another index for southern oscillation; but he also stated that the oscillation is dominated by an exchange of air between the South Pacific subtropical high and the Indonesian equatorial low, that it is generated spontaneously, and that its period varies roughly between one and five years and amounts to thirty months on the average. The period and time lag of the southern oscillation in his hypothesis is based on the advection of the anomalies of the sea surface temperature in the South Equatorial Current.

Ichiye (1966), using observational data from the equatorial Pacific between 140°W and

92°W, pointed out that strong trade winds produced upwelling and colder water and weak trade winds resulted in reduced upwelling and warmer sea surface temperatures. Krueger and Gray (1969) stressed that such colder water is believed to be primarily due to upwelling rather than to advection of cold water from the Peru Current and the South Equatorial Current.

Bjerknes (1966, 1969) analyzed the abnormally heavy rainfall in the so called equatorial Pacific dry zone during the period from 1950 to 1967, and then introduced a new circulation model, the "Walker Circulation," which is a longitudinal circulation model derived from the idea of Walker's southern oscillation.

In this paper, the author intends to show some relationships between the rainfall variations in the tropical and equatorial region from India to the equatorial Pacific and variations of low and middle latitude circulation in the Southern Hemisphere. The variability of the southern westerlies should be an alternative index for the atmospheric pressure of the South Pacific subtropical high in Walker's southern oscillation.

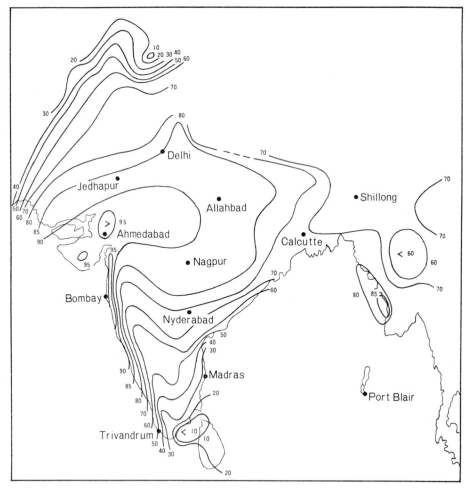

Fig. 2. Rainfall of June–September in India, expressed as percentage of annual (based on Rao, 1964).

2. RAINFALL VARIATIONS IN SOUTHEAST ASIA

a. *India and Vicinity*

As a result of detailed statistical analyses of much long-period data on rainfall in India, Pramanik and Jagannathan (1953) reported that there is no general tendency for increase or decrease in any particular area. However, year-to-year variations of annual rainfall are rather remarkable (Fig. 1).

In India, except for certain areas in the south and the extreme north, more than 75% of the annual rainfall occurs during the SW monsoon season (June–September) (Fig. 2); and as the variability of the NE monsoon rainfall is small (Rao, 1964), the variability of the SW monsoon rainfall is the main factor regulating year-to-year rainfall variations.

Although the sources of air masses or the main trajectories of air in the Indian summer monsoon are still unclear, the following observations can be made on the basis of a stream line analysis made on July 7, 1963, a typical summer monsoon day (Tsuchiya, 1970): 1) The Indian summer monsoon includes a remarkable flow from the Southern Hemisphere, and it is strongest within the lower troposphere (at about the 850 mb level). 2) The vertical extent of air current of the Indian summer monsoon less in the western part than the eastern part.

The SE trade winds include plentiful moisture in the lower levels during long sea travel over the Indian Ocean. Crossing the equator, the moisture extends to great heights and the air easily become unstable (Malurkar, 1960). It may be that stronger SE trade winds over the Indian Ocean are responsible for more rainfall over central India and its vicinity and vice versa, although other factors such as the mid-latitude westerly variations over the Northern Hemisphere, the thermal effect of the Tibetan plateau and many synoptic scale disturbances have an important effect on weather in India.

b. *Indonesia*

In Fig. 3 we can find a strong interdependence between the rainfall variations at Pon-

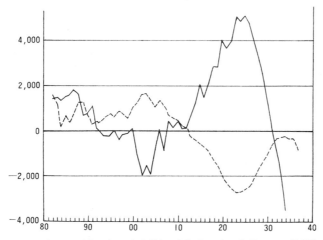

Fig. 3. 11-year moving total for the rainfall (mm) in Pontianak, Borneo (full line) and West Java (broken line) (based on ten Hoopen and Schmidt, 1951).

tianak, Borneo (a representative station along the equator) and those in West Java (average of twenty stations). This interdependence may be an outcome of the sonthward displacement of the ITC (intertropical convergence zone) in autumn and winter in the Northern Hemisphere, and of its northward displacement in spring; the higher pressure at Hong Kong in the overlapping 11–year totals for the air pressure is accompanied by a maximum of precipitation in West Java, and a minimum at Pontianak (ten Hoopen and Schmidt, 1951).

Concerning the influence of the intensity of the Siberian winter high, however, recently Sutrisno (1964) stated that an increase in rainfall over a certain period between the months of July and October is mostly the result of an increase in the frequency of tropical cyclones (or storms) in the western North Pacific and the Bay of Bengal, although statistical verification is not yet sufficient (Table I). It is better to say that the rainfall of Indonesia is very complicated matter, then it is difficult to show its general fluctuation or trend.

Table I. Rainfall at Djakarta and frequencies of tropical cyclones (storms) in the western North Pacific and the Bay of Bengal from July to October.

Year	Rainfall (mm) at Djakarta-Banten	Frequencies of tropical cyclones (storms)		
		N.W. Pacific	Bay	N.W. Pacific and Bay
1948	264	24	10	34
1949	224	17	4	21
1950	514	32	5	37
1951	396	12	9	21
1952	465	17	7	24
1953	92	15	5	20
1954	454	15	6	21
1955	632	20	8	28
1956	627	14	7	21
1957	367	14	5	19
1958	569	20	9	29
1959	224	16	8	24

Based on the data by Sutrisno (1964).

c. *Other Regions*

The characteristics of rainfall variations over a long period in other regions of Southeast Asia are also complex; however, it can be said that there are no stations showing a steady trend or stable long-term fluctuations (Fig. 4). Inter-annual variations at Manila and Pontianak show a relatively large amplitude, but the variations at Bangkok, Singapore, and Djakarta are less. Probably the greater amplitudes at Manila and Pontianak are due to the more effective Northern Hemispheric circulation. At Manila, typhoon frequencies are the main factors of year-to-year rainfall variations; at Pontianak, the complex interactions between the southern boundary of the Siberian high in winter and the ITC are the most probable causes.

3. RAINFALL VARIATIONS IN THE EQUATORIAL PACIFIC

a. *Eastern Part of the Equatorial Pacific*

Anomalous rainfall variations and the so called El Niño phenomena in the eastern part of the equatorial Pacific have a notable effect on air-sea interaction. Many authors have discussed the relationship between the sea surface in the Peru Current and rainfall variations

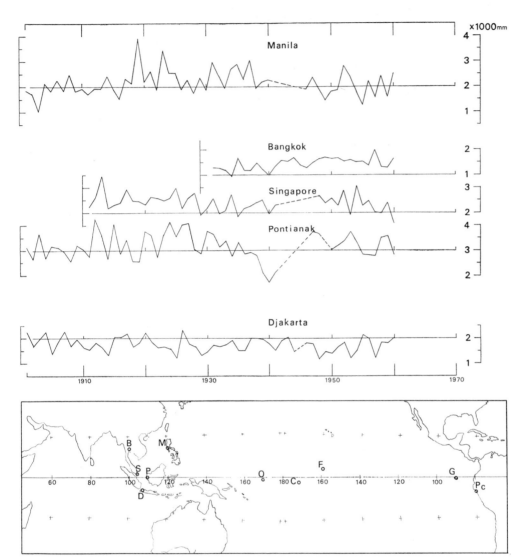

Fig. 4. Long-period rainfall (× 1,000 mm) variations at selected stations in Southeast Asia.

over the eastern part of the equatorial Pacific (Rodewald, 1958, 1964; Alpert, 1963; Schell, 1965; Berlage, 1966; Bjerknes, 1966; Doberitz, 1968; Tsuchiya, 1968ab).

During the well developed El Niño period (usually from December to March), very much higher sea water temperature advection prevails over the Peru Current as far south as Callao, Peru, and even into latitude 14°S, where prevailing upwelling during normal days disappears. Abnormally heavy rainfalls in the very dry area of the Peruvian coast occasionally occur during the El Niño periods. These phenomena extend as far west along the equator.

Figure 5 gives some examples of rainfall variations. Fluctuations in the Peru Current and upwelling are responsible for the El Niño phenomena, which generally depend on the eastern South Pacific (Wooster, 1961).

The origins of the El Niño sea water flow and Peru Current and their fluctuation mechan-

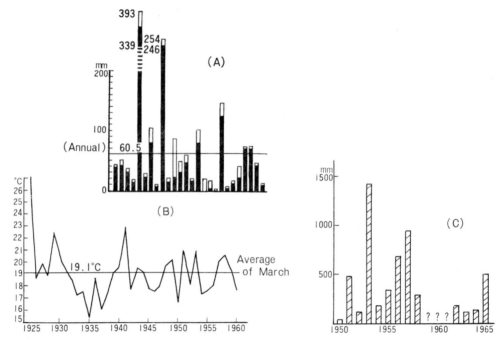

Fig. 5. Annual and 3 monthly (Feb.–April) rainfalls at El Alto, Peru (A) and monthly mean sea surface temperature (March) at Puerto Chicama, Peru (B) and annual rainfall at Puerto Baquerizo, Galapagos Isl. (C) (based on Tsuchiya, 1968a, b).

isms are not clearly understood; however, Schell (1965) proposed a hypothesis that "weaker westerlies and their marked convergence, linked to weaker southerlies and southeasterlies, are associated with higher sea surface temperatures along the Peru coast and with a weaker Peru Current and a weaker upwelling."

The relationship between the westerlies and low latitude air-sea interactions is discussed in a later section.

b. *Western Part of the Equatorial Pacific*

Canton Island, a small atoll located at 2°48′S and 171°43′W, represents the western conditions of the equatorial Pacific dry zone. The Canton Island type of rainfall regime is known to prevail along the equator from about 165°E eastward to the coast of South America, and a statistical study on the rainfalls of many stations in the dry zone revealed that the features of this area are a non-periodicity of annual rainfall and the existence of a remarkably coherent relationship among the island stations in this area (Doberitz, 1967, 1968).

The atmospheric and ocean conditions at Canton Island and its vicinity revealed some important air-sea interactions. Unusually big monthly totals of rainfall occur only during the periods when the ocean water is warmer than the atmosphere (Fig. 6). Under these conditions a maximum upward transfer of moisture takes place, and an intensive further convection brings the moisture to the level of condensation and up into towering shower clouds (Bjerknes, 1969).

Ocean Island (0°52′S, 169°35′E) lies to the west of Canton Island, and is situated in the

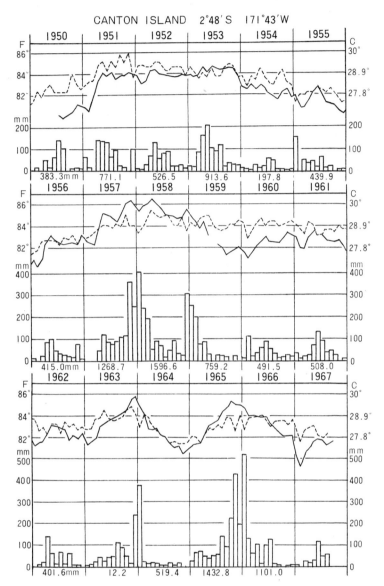

Fig. 6. Monthly air (broken line) and sea (full line) temperatures and monthly rainfall at Canton Island (Bjerknes, 1969).

western boundary of the dry zone. Fanning Island (3°51′N, 159°22′W) lies to the east of Canton Island, and the rainfall here shows features of the northern boundary of the dry zone. The annual cycle of rainfall at Fanning Island disappears or changes in wet or very dry periods (Fig. 7). At Ocean Island, we cannot find a stable annual cycle of rainfall during most observational years.

These three representative island stations show a notable pattern of rainfall over a long period, namely their anomalous wet or dry periods occurring at the same time.

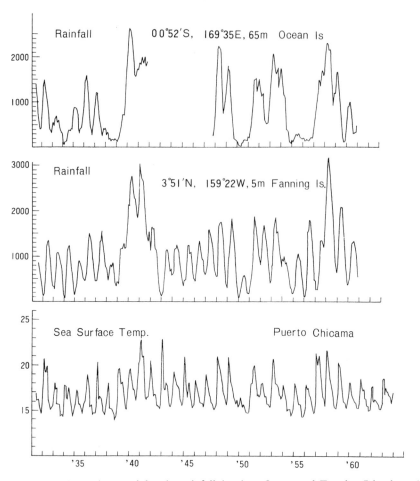

Fig. 7. 6-month moving total for the rainfall (mm) at Ocean and Fanning Islands and monthly mean sea surface temperature (°C) at Puerto Chicama, Peru.

4. LARGE AREAL ANALYSES ON THE YEAR-TO-YEAR VARIATIONS OF RAINFALL IN SOUTH-EAST ASIA AND THE EQUATORIAL PACIFIC

a. *Rainfall Variations*

Several maps constructed from "Die Witterung in Übersee," Hamburg Seewetteramt, revealed the remarkable reverse anomaly distributions on the annual rainfall amount in India and the equatorial Pacific. Their distributions were bounded by New Guinea and its vicinity as the nodal line (Fig. 8). In 1955, 1956, and 1962, the rainfall anomalies in the eastern region are negative and the anomalies in the western region are positive. In contrast, reversal types are predominant in 1957, 1958, 1965, and 1966, except for some irregularities in small areas. Ample rainfall in the eastern region and widespread drought conditions in the western region, particularly in central India and its vicinity, were observed.

These peculiar phenomena of rainfall over a wide range should be analyzed from the viewpoint of rainfall variability. Anomalous long period variations of rainfall in the equa-

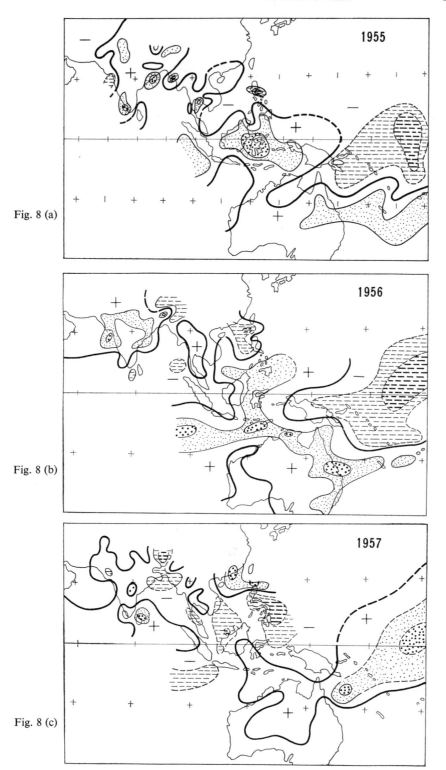

Fig. 8 (a)

Fig. 8 (b)

Fig. 8 (c)

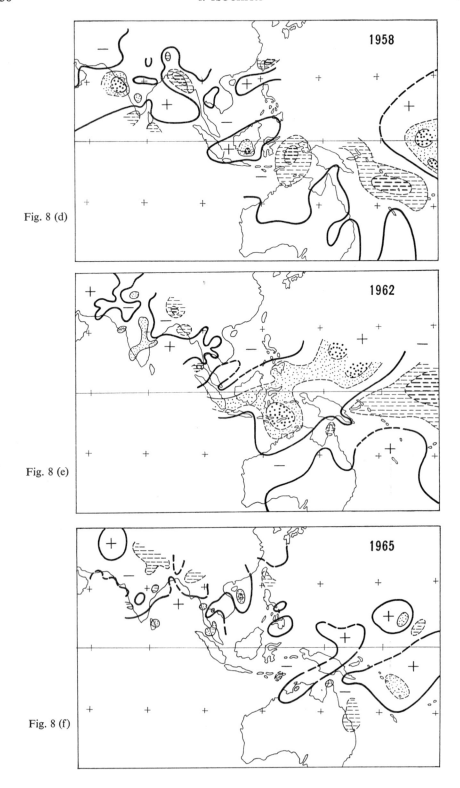

Fig. 8 (d)

Fig. 8 (e)

Fig. 8 (f)

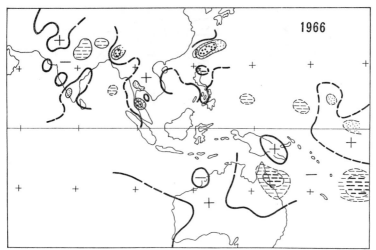

Fig. 8 (g)

Fig. 8a–g. Annual rainfall anomaly distributions over the India—equatorial Pacific region (:: :> +800mm, :: :> +400mm, ≡ :< −400mm ≡ :< −800mm; constructed from "Die Witterung in Übersee").

torial Pacific dry zone are discussed in a number of papers (Ichiye and Petersen, 1963; Bjerknes, 1966, 1969; Doberitz, 1968; Krueger and Gray, 1969). Areal rainfall anomaly analyses during the Indian summer monsoon months (June–September) over India have been reported annually since 1959 (*Indian J. Met. Geophys.* 1960–1968). However, more extended global analyses in the region from India to the equatorial Pacific remain to be undertaken.

In general, the equatorial Pacific dry zone extends from the coast of Peru to 165°W. Within this zone, the annual rainfall averages less than 800 mm, but the zone extends to 165° E or more westward in rainless years such as 1955, 1956, and 1962, and disappears in very wet years such as 1957, 1958, 1965, and 1966.

It can be said that the reverse anomaly distributions mentioned above occurred also before 1955, although their areal analyses were rather difficult. According to the Indian droughts and floods data (Ramdas, 1960) and the secular variations of two representative stations in the dry zone (see Fig. 7), there were patterns of droughts in India—plentiful rainfall in the dry zone in 1939, 1941, 1951, and 1952, and reversal types in 1942 and 1944.

b. *The Southern Oscillation*

The southern oscillation is a complex of interactions over a very large area. According to Walker and Bliss (1932): "In general terms, when pressure is high in the Pacific Ocean, it tends to be low in the Indian Ocean from Africa to Australia; these conditions are associated with low temperatures in both these areas, and rainfall varies in the opposite direction to pressure." Although the southern oscillation index or Walker's index is a composite index based on several meteorological factors over the India–Pacific region, its value is high when pressure is high over the Pacific and low over the Indian Ocean, and this convention is generally opposite in sign to that used by Berlage, whose index is the pressure at Djakarta, Indonesia (Troup, 1965).

An example of the southern oscillation (Berlage, 1957) shows the worldwide distribution of correlations of annual pressure anonalies with simultaneous pressure anomalies in

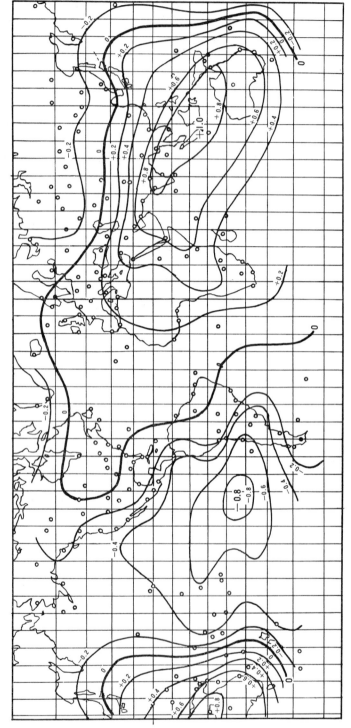

Fig. 9. Walker's southern oscillation as presented by Berlage (1957). (The map shows the worldwide distribution of correlations of annual pressure anomalies with simultaneous pressure anoamalies in Djakarta, Indonesia).

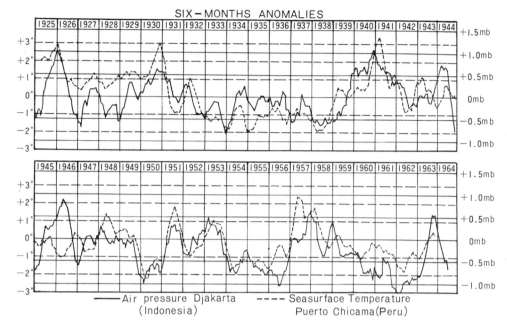

Fig. 10. Running 6-month average anomalies of Djakarta air pressure and Puerto Chicama sea surface temperature (Berlage, 1966).

Djakarta (Fig. 9). Berlage (1966) indicated a fine parallel long-period variation in Djakarta pressure and sea surface temperature at Puerto Chicama, Peru, since 1925 (Fig. 10).

The relationship between the southern oscillation and the rainfall variation was a confusing matter until quite recently. However, the parallels between the sea surface temperature and rainfall at Canton Island (Fig. 6) and the rainfall at Ocean and Fanning Islands and the sea surface temperature at Puerto Chicama (Fig. 7) are remarkable. Therefore, they suggest a reversal phenomena on the rainfall variations in India. These large areal analyses reveal the possibility of a relationship between the southern oscillation and the rainfall variation in India and the equatorial Pacific.

c. *The Walker Circulation*

According to Bjerknes (1969), when the cold water belt along the equator is well developed, the air above it will be too cold and heavy to join the ascending motion in the Hadley circulation. Instead, the equatorial air flows westward between the Hadley circulations of the two hemispheres to the warm western Pacific. There, after having been heated and supplied by moisture from the warm waters, the equatorial air can take part in large-scale, moist-adiabatic ascent (Fig. 11).

Bjerknes named this new circulation model the "Walker circulation" since it can be shown to be an important part of Walker's southern oscillation. He also indicated that when the Walker circulation axis is located east of 172°W, Canton Island is under a rising air column with frequent rain from midtropospheric clouds as well as from convective clouds favored by the positive sea-minus-air temperature difference, and suggested as a mechanism the weak thermally driven air circulation along the equator with sinking air over Africa and rising air over Indonesia.

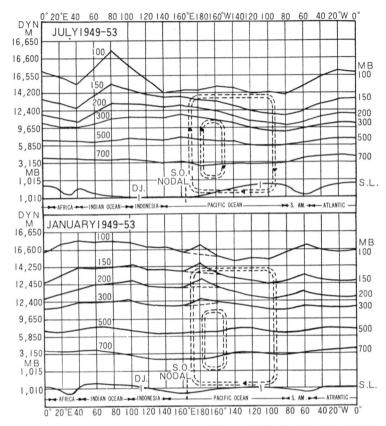

Fig. 11. Walker circulation over the Pacific (Bjerknes, 1969). Profile of height (dynamic meter) of standard isobaric surfaces along the Equator in January and July.

Thus, the fluctuation of the Walker circulation is responsible for initiating some of the major pulses of the southern oscillation. Weaker or stronger developments and eastward or westward axis movements in the Walker circulation can be referred to some widescope rainfall variations. For example, weak Walker ciruclation is responsible for conditions during the years of 1957–1958 and 1965–1966. The normal or relatively stronger Walker circulation is related to conditions during the years of 1962 and probably 1967–1968.

5. SE TRADE WIND VARIATIONS AND THEIR INFLUENCES

A fine example of air-sea interactions is recognized along the equatorial Pacific dry zone. Prevailing easterlies advect cold water from the South Equatorial Current and strong trade winds cause upwelling (a process of vertical water motion in the sea, where sub-surface water moves toward the surface) and colder water, while weak trade winds result in reduced upwelling and warmer ocean temperatures (Ichiye, 1966). The observational wind data from surface to air and oceanographic data in the equatorial Pacific are insufficient to verify these phenomena; however, it is possible to assume that the year-to-year variation of SE trade winds which is the main driving force to the Peru Current and the South Equatorial Current affects the sea surface temperature in the dry zone through the year-to-year varia-

tion of upwelling. The extinction of lower tropospheric easterlies in anomalous heavy rainfall periods and the strong upwelling and dry weather in the prevailing easterlies (Ichiye and
Petersen, 1963; Ichiye, 1966; Bjerknes, 1969) are evidence of these air-sea interactions.

For example, in a case study of the anomalous rainfall of the 1957–58 winter, Ichiye and
Petersen (1963) reported that the daily wind data at Canton Island revealed that the wind
direction shifted suddenly from the usual easterly trade winds to westerlies on November 17
and continued to be almost entirely westerly unitl December 9. Using the observational
oceanographic and wind data in the equatorial Pacific between 140°W and 92°W, Ichiye
(1966) also showed the upwelling due to the effect of the westward wind stress and suggested
the role of the SE trade winds variation.

The monthly zonal wind component from surface to air at Canton Island for January 1966
and January 1967 shows a clear wind shift: the profile in 1966 shows a weak wind or westerly
component while the profile in 1967 shows a strong easterly component at the lower troposphere and a strong westerly component at the upper troposphere (Krueger and Gray,
1969). Therefore Bjerknes's Walker circulation was much more intense in January 1967 than
during the previous year.

It is difficult to make any analyses for the Indian Ocean since the observational data are
very few. It will be necessary to wait for global video observations of th earth's clouds from
meteorological satellites instead of real wind data.

6. Variations of Mid-Latitude Westerlies in the Southern Hemisphere and Their Influences

Schell (1965) states that weaker southerlies and southeasterlies during the March–November period are associated with higher sea surface temperatures during the following December–February period in the southeastern Pacific. The scheme of weaker westerlies to weaker

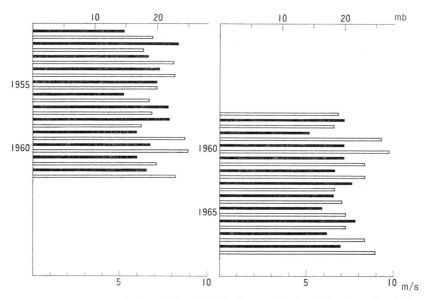

Fig. 12. Year-to-year variations of the mid-latitude westerlies in the Southern Hemisphere
(■: January, ☐: July; Data corresponds to Table II).

southeasterlies (SE trade winds) and stronger westerlies to stronger southeasterlies is a probable expression although the lag phenomena mentioned above are not clear and many mechanisms relating to this scheme remain unsolved.

The actual state of the mid-latitude westerlies in the Southern Hemisphere is not clear at present except in some limited regions. Many authors have included the mean southern westerlies in their articles (van Loon, 1964, etc.). Before the IGY period, the data for high latitudes in the Southern Hemisphere were very poor, especially in the southern winter; therefore these mean westerlies are rather questionable.

The author computed the sea level wind speed at 45°S from the pressure difference between 35°S and 55°S in January (southern summer) and July (southern winter) based on monthly mean charts for successive years by Lamb and Johnson (1966) and also by the Hamburg Seewetteramt (1957–1968). Fig. 12 shows the year-to-year variations of the hemispheric mean westerly indices (zonal indices), and Fig. 13 shows the same factors for four sectors in the Southern Hemisphere.

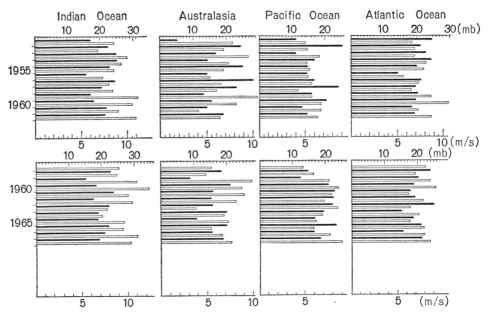

Fig. 13. Year-to-year variations of the mid-latitude westerlies in the Southern Hemisphere by four sectors (notations are the same as in Fig. 12).

The variations of these zonal indices are remarkable, and there are several years in which the southern summer westerlies are stronger than the southern winter westerlies although the mean values are higher in winter (Table II).

Remarkable abnormalities of the westerlies prevailed over most of the sectors in 1957 and 1958. Winter (July) indices are very low and their indices are lower than summer (January) in most sectors. These features repeatedly occurred in the two periods centered in 1963 and 1966. Abnormal rainfall events at Canton Island occurred during the periods of 1963–1964 and 1965–1966 (see Fig. 6). Widespread drought conditions are notable features in the sum-

Table II. Average zonal westerly indices at sea level (35–55° latitude; m/s).

	Indian Ocean	Australasia	Pacific	Atlantic	S. Hemisphere	N. Hemisphere
Jan.	9.1	5.8	5.1	8.2	7.1	2.6
July	9.2	5.9	5.7	6.5	6.8	1.1
			After van Loon, 1964. (Jan. 1951–57, July 1951–56)			
Jan.	7.5	6.3	6.0	7.4	6.8(6.2)	(2.0)
July	9.1	7.2	5.9	7.6	7.5(6.5)	(1.1)
			Computed from the data of Lamb and Johnson (1966) (1951–62)			
Jan.	7.2	5.9	6.9	7.1	6.8	
July	9.9	7.2	7.5	7.6	7.9	
			Computed from the data of the Hamburg Seewetteramt			
			(Jan. 1958–68, July 1957–68)			

Indices in parentheses are computed from normal tables by the Forecasting Research Lab., Meteorological Research Institute, Tokyo (1968).

mer monsoon seasons of 1957, 1958, 1965, and 1966 over India and its vicinity (see Fig. 8). The rainfall distribution patterns in 1967 and 1968 over the region from India to the equatorial Pacific were normal, namely wetter in the west and drier in the east, and the southern westerly indices were higher in July while the annual averages (means of July and January indices) were rather high. These facts do not contradict the statement by Schell (1965) except that several expressions include a lag mechanism (see the first part of this section). His hypothesis is as follows: "Weaker westerlies and their marked convergence, linked to weaker southerlies and southeasterlies, are associated with higher sea surface tmperature along the coast (Peru coast) and with a weaker Peru current and a weaker upwelling."

Previous statements indicate the possibility of a working model illustrating a relation between the southern westerlies and the rainfall over the region from India to the equatorial Pacific, although the mechanisms by which the southern westerlies affect the SE trade winds in the South Pacific and the Indian Ocean are as yet unsolved both analytically and theoretically. The following diagram can be presented, however, as a hypothetical model for periods of normal or relatively strong westerlies.

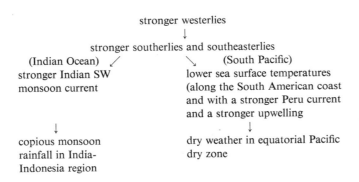

This model is reversed during the period of weaker westerlies. In other words, the weaker southern westerlies provide less rainfall in India and its vicinity then plentiful rainfall in the equatorial Pacific dry zone. The rainfall data in the region from India to the equatorial Pacific and the sea surface temperature at Puerto Chicama suggest that there are shorter time lags in these processes. We can assume that a stronger Walker circulation (Bjerknes, 1969) coincides with a situation of stronger westerlies and vice versa.

There are many problems to be solved in this model and the relation between the wester-lies and the SE trade winds is particularly difficult, because the meteorological data for these two oceans are insufficient. However, cloud data and calculated wind data based on cloud motions obtained from satellite cloud photographs will be available in the near future for solving these questions.

7. Concluding Remarks

As a result of investigating much data, including rainfall, droughts and floods, atmospheric pressure and sea surface temperature in India, the equatorial Pacific and the Southern Hemi-sphere, several characteristic features connected with the year-to-year variations in general circulation have been clarified. The southern oscillation theory remains a problem to be analyzed from the viewpoint of the wide scope of air-sea interactions and general circulation mechanisms in the Southern Hemisphere.

Inter-annual and seasonal variations of mid-latitude westerlies in the Southern Hemi-sphere are remarkable. There are several cases in which summer (January) westerlies are stronger than winter (July) westerlies, although long period average westerlies are stronger in winter. Abnormal distributions in annual rainfall occurred, especially in the southern winter, in years of weak westerlies such as the years of 1957, 1958, 1965, and 1966, in which the patterns of the western region in the east and the drier region in the west are recognized. Similar phenomena probably occurred in 1940–1941 and 1951–1952. In these years higher sea surface temperatures at Puerto Chicama are recognizable (see Fig. 7). A drier east and wetter west is a pattern which results when normal weather prevails in the India—equatorial Pacific region, and it occurred in the year of stronger westerlies, especially in the southern winter.

Owing to the scarcity of observational wind data from surface to air in the regions in question and to the difficulty of areal analysis of rainfall which is subject to complexed mechanisms in the western region, there is little likelyhood of finding a simple relationship between areal rainfall anomaly and variability of westerlies. However, Indian drought and flood data show that there is no evidence for a widespread drought condition in India oc-curring during the period of less rainfall in the equatorial Pacific dry zone and lower sea surface temperature at Puerto Chicama. It may therefore be possible to study the mechanism of the areal rainfall variations in the India–Indonesia region through variations in the SE trade winds using the variability of the southern westerlies as well as cases in the equatorial Pacific dry zone.

References

Alpert, L. 1963: The climate of the Galapagos Islands. *Occ. Papers, Calif. Acad. Sci.* **44** 21–44.
Berlage, H. P. 1957: Fluctuations of the general atmospheric circulation of more than one year; their nature and prognostic value. *Meded. Verhand., Konink. Nederlands Met. Inst.* **69** 152p.
Berlage, H. P. 1966: The southern oscillation and world weather. *Ibid.* **88** 152p.
Bjerknes, J. 1966: A possible response of the atmospheric Hadley circulation to equatorial anomalies of ocean temperature. *Tellus* **18** 820–829.
Bjerknes, J. 1969: Atmospheric teleconnections from the equatorial Pacific. *Month. Weather Review* **97** 163–172.

Chen, Shi-shun, Chen, Lian-bau 1965: The annual variation of the precipitation and their characteristics during recent sixty years in southern China. *Acta. Geogr. Sinica* **31** 304–320 (in Chinese).

Doberitz, R. 1967: Teleconnections and phase relations of rainfall at the tropical Pacific Ocean. Final Tech. Rep. Contract No. DA-91-591-EUC-3983, European Research Office, U.S. Army, Univ. Bonn, 39p.

Doberitz, R. 1968: Cross-spectrum analysis of rainfall and sea temperature at the equatorial Pacific Ocean—A contribution to the "El Niño" phenomenon. *Bonner Met. Abhand.* (8) 61p.

Forecasting Research Laboratory, Meteorological Research Institute, Tokyo, 1968: Normals of monthly mean sea-level pressure for the Northern and Southern Hemispheres. *Tech. Rep. Japan Met. Agency* Tokyo (61) 171p.

Hamburg Seewetteramt, 1955–1968: Die Witterung in Übersee (1955–1968).

Ichiye, T. 1966: Vertical currents in the equatorial Pacific Ocean. *J. Oceanogr. Soc. Japan* **22** 274–284.

Ichiye, T., Petersen, J. R. 1963: The anomalous rainfall of the 1957–58 winter in the Equatorial Central Pacific dry area. *J. Met. Soc. Japan* **41** 172–182.

Indian J. Met. Geophys. 1960–1968: Rainfall and floods during (1957–1967) southwest monsoon period. **11** 215–222, **12** 173–181, **13** 147–156, **14** 253–260, **15** 337–346, **16** 331–340, **17** 327–332, **18** 329–334, **19** 267–272.

Krueger, A. F., Gray, Jr., T. I. 1969: Long-term variations in equatorial circulation and rainfall. *Month. Weather Review* **97** 700–711.

Lamb, H. H., Johnson, A. I. 1966: Secular variations of the atmospheric circulation since 1750. *Geophys. Mem.* (110) 125p.

Malurkar, S. L. 1960: Monsoons of the world—Indian monsoons. *In* Monsoons of the world, India Met. Department, 92–99.

Pramanik, S. K., Jagannathan, P. 1953: Climatic changes in India—(1) rainfall. *Indian J. Met. Geophys.* **4** 291–309.

Ramage, C. S. 1969: Summer drought over western India. *Yearbook, Assoc. Pacific Coast Geographers* **30** 41–54.

Ramdas, L. A. 1960: The establishment, fluctuations and retreat of the Southwest Monsoon of India. *In* Monsoons of the world, India Met. Department 251–256.

Rao, K. N. 1964: Seasonal forecasting—India. *WMO Tech. Note* (66) 17–30.

Rodewald, M. 1958, 1964: Beiträge zur Klimaschwankung in Meere (1), (13). *Deutsche Hydrogr. Z.* **11** 78–82, **17** 105–114.

Schell, I. I. 1965: The origin and possible prediction of fluctuations in the Peru Current and upwelling· *J. Geophys. Res.* **70** 5529–5540.

Sutrisno, C. 1964: The effect of tropical cyclones in the western North Pacific and the Bay of Bengal on the weather in the southwestern part of Indonesia. *Proc. Symp. Tropical Met.* 1963, New Zealand 201–206.

ten Hoopen, K. J., Schmidt, F. H. 1951: Recent climatic variation in Indonesia. *Nature* **168** 428–429.

Troup, A. J. 1965: The southern oscillation. *Quart. J. Roy. Met. Soc.* **91** 490–506.

Tsuchiya, I. 1968a: El Niño at Peru coast, South America (in Japanese). *Chiri* (Geography) Tokyo **13**(2) 93–97.

Tsuchiya, I. 1968b: The rain at Galapagos Islands (in Japanese). *Ibid.* **13**(4) 73–78.

Tsuchiya, I. 1970: Year-to-year variations of rainfall over the India—equatorial Pacific region and low and middle latitude circulations in the Southern Hemisphere. *Pap. Met. Geophys.* **21** 73–87.

van Loon, H. 1964: Mid-season average zonal winds at sea level and at 500 mb south of 25 degrees, and a brief comparison with the Northern Hemisphere. *J. Appl. Met.* **3** 554–563.

Walker, G. T. 1924: Correlation in seasonal variations of weather; Pt. 9, A further study of world weather. *Mem. Ind. Met. Dept.* **24** (Pt. 9) 275–332.

Walker, G. T., Bliss, E. W. 1932: World weather (5). *Mem. Roy. Met. Soc.* **4**(36) 53–84.

Wooster, W. S. 1961: Yearly changes in the Peru Current. *Limnol. Oceanogr.* **6** 222–226.

CLIMATIC CHANGE IN THE QUATERNARY IN ASIA: A REVIEW

Hiroshi TABUCHI AND Kazuko URUSHIBARA

Abstract: The purpose of this paper is to review the climatic changes in the Quarternary in Southeast Asia. The writers have used the following factors as parameters: glaciers, pluvial lakes, coral reefs, snow line, forest line and permafrost. The Climatic Optimum around 6,000 B.P. and Climatic Deterioration around 2,500 B.P. occurred simultaneously in Monsoon Asia. It was arid in Central Asia, and humid in the Near East and West Pakistan around 6,000 B.P. Around 2,500 B.P., it was arid in central Asia and West Pakistan and humid in the Near East and Japan.

These circumstances are thought to have occurred because of the strong Southwest Asian polar frontal zone over the Near East and West Pakistan in winter around 6,000 B.P., in contrast to the weak frontal activity in Central Asia. Another factor may have been intensification of the anticyclone over the Eurasian Continent in winter around 2,500 B.P., indicating strong frontal activity of the polar frontal zones in the Near East and over Japan, at the margin of the anticyclone.

1. INTRODUCTION

The purpose of this paper is to review climatic changes from the Pleistocene to the Recent period in Southeast Asia. Because of insufficient data concerning climatic changes in this region, there is little information on the subject at present. There are many papers on West Asia, Siberia and East Asia, but only a few on Southeast Asia. It is well known that the snow line, forest line, permafrost, coral reefs and other factors are useful parameters for climatic change. Therefore, we describe these parameters in Southeast Asia during the period under discussion. The first part of this paper was written by K. Urushibara and the second part by H. Tabuchi.

2. PREVIOUS STUDIES

a. *General Views*

Quaternary climatic changes on the European and North American continents have been studied intensively in recent years. Pleistocene climatic changes in particular have been clarified by means of C_{14} datings, varve clay, pollen analysis and annual rings. In central and northern Europe, the Late glacial period (prior to the Postglacial, 15,000 B.P.–10,000 B.P.) was broken down into three periods, cold, warm and cold. Then in the last years of the Late glacial period, the temperature rose abruptly. The Post glacial period has been classified as follows: the Boreal period (becoming warm), the Atlantic period (Climatic Optimum) the Sub-Boreal period (cold in winter) and the Sub-Atlantic period (colder).

AROUND 20,000 YEARS B.P.

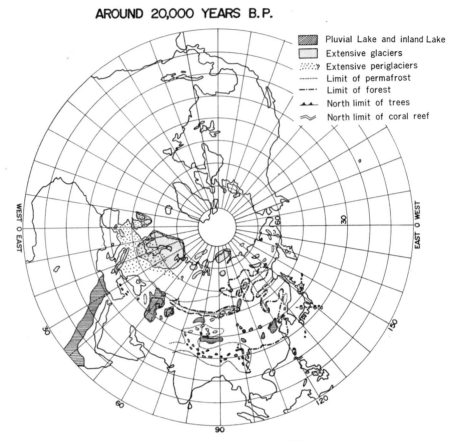

Fig. 1. Conditions in Asia around 20,000 years B. P.

On the North American continent, climatic changes in the Late glacial period are not yet as well defined as on the European continent, and there are many opinions concerning the division of the Late glacial period. There is some disagreement on the existence of the three subperiods: the cold, the warm and the cold. In the Postglacial period, climatic changes were similar to those on the European continent.

It is interesting to note that on both continents, the time of Climatic Optimum occurred simultaneously. Inland in North America, the Great Basin region was arid, but Southern Canada was humid. On the other hand, Northern Europe was humid, in contrast to the arid area around the Black Sea. Generally speaking, it seems that the inland areas in the middle latitude were warm and arid during this period.

There are only a few reports on climatic changes in Monsoon Asia, and it is therefore very difficult to compare climatic changes in this area with those on the European and North American continents. However, as a general outline, charts have been prepared on the basis of earlier studies of climatic changes since the Würm ice age in Monsoon Asia.

Figures 1–4 show the four periods during which remarkable climatic changes occurred: 1) the Würm ice age, 2) around 6,000 B.P. (Climatic Optimum), 3) around 4,000 B.P. (sub-Boreal period), 4) around 2,500 B.P. (onset of the Sub-Atlantic period).

AROUND 6,000 YEARS B.P. (CLIMATIC OPTIMUM)

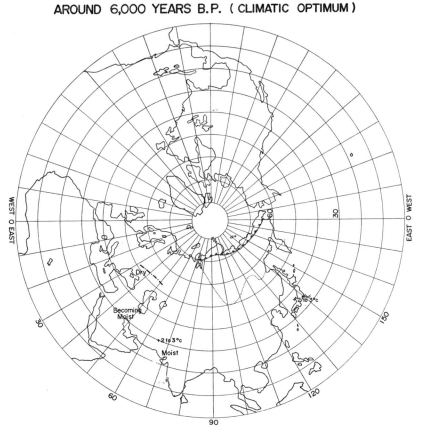

Fig. 2.   Conditions in Asia around 6,000 years B. P. (modified from Lamb, 1966).

Figure 1 is based chiefly on results obtained by Wissmann (1938), Frenzel and Troll (1952), with the data on Japan contributed by Kobayashi and Hoshiai (1955), Kaizuka (1970) and Suzuki (1970). Figures 2, 3 and 4 were compiled only for Asia mainly from results obtained by Lamb *et al.* (1966) and from data on Japan obtained by Fuji (1965, 1966), Kaizuka (1964) etc.

The Climatic Optimum in almost the entire area of Monsoon Asia apparently corresponded to that on the European and North American continents. The areas in inland Asia were arid and warm, and West Pakistan and the Near East were humid and warm. On the other hand, the area on the northern coast of the Black Sea was arid. The evidence indicates that the arid and warm conditions in the Climatic Optimum in these areas in inland Asia coincide perfectly with conditions in inland Europe and North America at middle latitudes. In the Sub-Boreal period, however, West Pakistan was arid and Central Asia was relatively humid. At the beginning of the Sub-Atlantic period, inland Asia was also arid, but, in contrast, the Near East, Southwest Asia and Japan were humid.

These climatic chnges can be attributed to changes in atmospheric circulation patterns and transitions in the frontal zones. The mechanism of these climatic changes cannot be dealt with in detail in this paper, however, because of the heterogeneous and regionally insufficient data for the areas studied.

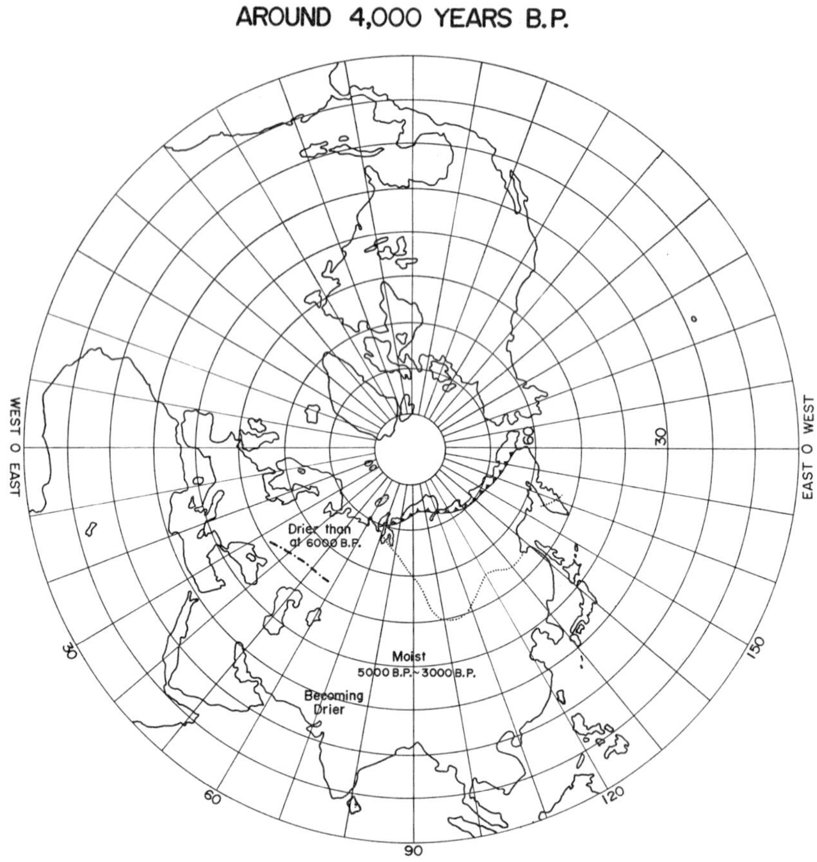

AROUND 4,000 YEARS B.P.

Fig. 3. Conditions in Asia around 4,000 years B. P. (modified from Lamb. 1966).

b. *Climatic Changes in Japan*

The snow lines during the Würm ice age revealed geomorphologically by cirques and other periglacial morphology are lower than the levels calculated theoretically for the present time: they are 1,400m in the Japanese Alps and 1,800m lower in the Hidaka mountains. The temperature at the time was 5°–6°C lower than at present. During the early period of recent geological time after the Würm ice age, changes in the vegetation and climate are not very clear.

Studies in postglacial chronology based on pollen analyses were begun in Japan around the 1950's, and climatic changes have since become clearer. On the basis of pollen analyses, Nakamura (1967) classified the Recent period as follows:

L, RI: (12,000–9,000 B.P.) The forest zone was 1,000m lower than at present.

RII: (9,000 B.P.—3,000~4,000 B.P.) Became warm. The forest zone was 200–500m higher than at present.

RIIIa: (3,000~4,000–1,500 B.P.) Became cold. Increased rainfall. The forest zone was almost the same as at present.

AROUND 2,500 YEARS B. P.

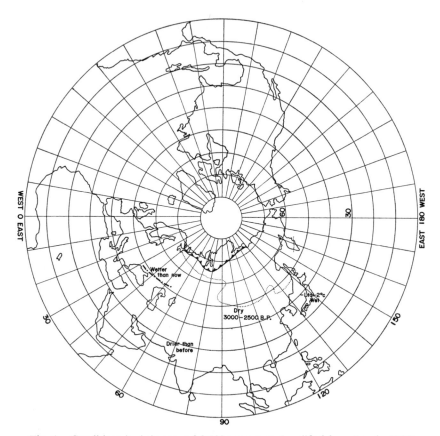

Fig. 4. Conditions in Asia around 2,500 years B. P. (modified from Lamb, 1966).

RIIIb: (1,500 B.P.–present)

Nakamura, in an earlier paper (1952), drew the boundary between the periods of RII and RIII, as just before the maximum occurrence of conifers. On the other hand, Tsukada (1958) concluded that the boundary should be set at an earlier period, when the conifers were beginning to increase.

From geomorphological studies on marine terraces and pollen analyses, the relation between changes in sea level and climatic changes on the Pacific coast of Japan was made clear (Sakaguchi, 1963; Kaizuka, 1964). According to their findings, the mean annual temperature aorund 6,000 B.P. (Numa stage) was 2°C higher than at present. The sea level was several meters higher, corresponding to the Climatic Optimum in Europe. Around 4,000 B.P. to 2,000 B.P., the sea level reached its highest, and was at about five meters. After that it began to retreat gradually. According to the results of pollen analyses in Oze, there was a low temperature period around 2,500 B.P., which corresponded to the Climatic Deterioration in Europe. From investigations of lake terraces and C_{14} dating of buried wood in various parts of Japan, it was concluded that it became cold in 3,000–1,500 B. P. and that at the same time the precipitation increased, indicating a humid climate (Horie, 1964). During this period, however, it was rather arid in inland Asia, so that it can be supposed that the relatively apparent frontal zones existed over East Asia.

On the Japan Sea side of Japan, difference of change in the sea level uplift and the crustal movement reached around 6,000 B.P. to a maximum of 6m, and the temperature rose 2°C. Around 4,000–1,500 B.P. (late Jômon and Yayoi periods), the sea level was 2m lower than at present and the mean annual temperature fell to −1 to −2°C (Fuji, 1965).

Climatic changes in historical times will be dealt with next. Studies on secular changes in climate revealed from annual rings at Mt. Arisan in Formosa and at Mt. Kiso in Central Japan, the numbers of freezing days at Lake Suwa in Central Japan, and the first flowering date for cherry blossoms in Kyoto, have led to the conclusion that a 700-year cycle can be observed in climatic history (Nishioka, 1949). 89 to 100–year cycles, as well as the 35-year cycle suggested by Brückner were also found. On the other hand, Taguchi (1939–1942), on the basis of the first flowering date for cherry blossoms and data on droughts and storms insisted that there were no remarkable climatic cycles; and considering anomalous periods of protracted cold weather in summer, no long cycles over several hundred years.

According to previous studies, climatic changes in historical times were as follows: In the 5th to 6th centuries the temperatures were low, and then from the end of the 7th to the beginning of the 8th century, it was cold and humid in summer. In the 8th to 9th centuries, it was warm and dry, resulting in the early flowering date for cherry blossoms in Kyoto. Yamamoto is of the opinion that it was arid in Southwest Japan, under the influence of the subtropical air mass or subtropical anticyclone over the Northwest Pacific, and therefore much precipitation was observed in northern Japan, brought by the active frontal zone on the northern margin of the subtropical air mass (Yamamoto, 1967). From the 11th to 14th and 15th to 16th centuries it was cold and humid, and in the 19th century it grew cold and humid again, so that the first flowering date for the cherry blossoms in Kyoto was late. Yamamoto (1967) states that the precipitation from the 19th century to the present shows an inverse correlation in Southwest Japan and the Hokkaido district. This relation can also be applied to the historical period. Therefore climatic changes in Japan can be seen in relation to those in East Asia. Around the 15th century it was cold and humid in Southwest Japan, but it is said that Korea was arid in summer. In the small ice age at the beginning of the 19th century it was humid on the Pacific side of Japan. During the same period, however, it was arid in Korea. These conditions resulted from the activity of frontal zones running along the Pacific coast of Southwest Japan, because of the southward shift of the frontal zones between the strong Okhotsk air mass (Pm air mass) and the weak subtropical air mass (Tm air mass) (Yamamoto, 1967).

In the 15th century, China had an arid climate similar to Korea's. Prior to this period, however, no clear relation can be observed between China, Korea, and Japan. With respect to temperature conditions, a warm period occurred in the 8th to 9th centuries in both the eastern part of China and Japan and a cold period in the 19th century in Korea and Japan and probably also in China. The chief climatic changes in Asia are indicated in Table I.

    c. *Climatic Changes in Eastern and Central Asia*
    1) Korea
Climatic changes during the Recent period in Korea have been clarified by means of pollen analyses (Matusima, 1941; Yamazaki, 1940). It was found that Alnus and Quercus were dominant in one period and red pine trees (Pinus densiflora) in the next. These periods corresponded to the end of RII or the beginning of RIII in Japan.

Table I. Climatic change of postglacial age in Asia.

B.P.	Europe	Western Asia — Near East	Western Asia — West Pakistan	South Asia	Central Asia including China	Eastern China including Manchuria	Eastern Asia — Korea	Eastern Asia — (Pacific side)	Eastern Asia — Japan	Eastern Asia — (Japan sea side)
0		precipitation increased to some extent			14c increase of rainfall in Asia	humid at present (once, there occurred dry period) 15c arid	19c dry cold 15c arid	sea level slow fall	humid 19c cold humid 15c cold (Southwest Japan)	
500					dry					
1000	Sub-Atlantic	590–645 A.D. dry		600–1200 A.D. dry (Angkor Wat)	7–8c dry, drought many ruins were abandoned	8–9c warm			dry 8–9c warm	
2000		humid warm in winter	drier than before	increase of rainfall in Southwest Asia	2c many ruins were abandoned			high } cold	increase of rainfall	−1~2°C −2m sea level fall
3000	Sub-Boreal	decrease of rainfall	becoming dry		dry extension of glaciers		3000 ~ 4000 B.P.	level		
4000										
5000								slow rise		
6000	Atlantic Climatic Optimum	increase of rainfall +2~3°C humid	+2~3°C humid		dry and warm mountain glaciers disappeared at the middle latitudes			+2°C 6000 B.P. sea level rose several meters	sea level rose 6m (max.)	
7000								remarkable rate of rise		
8000	Boreal	increase of rainfall								
9000	Pre-Boreal	became warm								
10000										

Eastern Asia zones: RIIIb, RIIa, RII warm, RI cold

In historical times, the most recent climatic changes were clarified from data of rainfall in Seoul. From 1770 to 1933 A.D., striking cycles of 2, 6, 13, 35 and 37 years were observed (Tada, 1938). Periods of drought were also noted around 1,400 A.D. and at the beginning of the 19th century (Yamamoto, 1967). There was a dry period around 1,900 A.D. (Hoyanagi, 1961), while an arid period was also observed in the 19th century in Manchuria, northern China and Mongolia.

2) China

The classification of the ice age in the mountainous areas of China and its relation to the ice age in Europe are not clear due to the scarcity of studies, but in the Quaternary, at least three to four ice ages occurred.

In the last ice age, the snow line on the Tibetan Plateau and on the Himalayan mountains became lower. In the central part of the Tibetan Plateau, however, due to the small amount of precipitation, the snow line changed only a little, that is, it stayed at a higher level than at the southeastern part. The present glacier in the Tienshan mountains developed during the Little ice age in the postglacial age (Wissmann, 1959, 1960; Hoyanagi, 1966). The climatic changes during the postglacial age in China are vague, but it is known that Central Asia became dry around 6,000 B.P., and that the high mountain glacier disappeared. The glacier expanded and the lake level reached its maximum at the onset of the Sub-Atlantic period. By contrast, Central Asia grew arid at 3,000 to 2,500 B.P., although in northern Manchuria, it is at present humid. According to a study by Asakura published in this volume, a moist region exists in Manchuria, whose source of moisture may be in western Siberia. On the other hand, it is known geomorphologically that there was an arid period in Manchuria (Tada, 1950, 1964), even though its absolute time is unknown. Presumably, therefore, when western Siberia did not provide a marked source of moisture because of its colder climate, no moist region appeared in Manchuria, that is, it was arid there.

Climatic changes in historical times are fairly well known from ancient manuscripts, ruins etc. According to a paper by Chu, Co-Ching (1954), studies were made by Alexander Hosie of the frequency of occurrence of floods and droughts in Chinese chronicles from the Tang to the Ming dynasties: It was apparently extremely dry from 300 to 400 A.D. and dry from 500 to 700 A.D. From 1,100 to 1,200 A.D. and from 1,300 to 1,400 A.D. it was humid, and then it again became dry from 1,400 to 1,500 A.D. On the other hand, an arid period is known to have occurred from 600 to 1,200 A.D. in Central Asia (Brooks 1926, Willet 1949, Chu 1954). Brooks (1926) states that in China there was an arid period from the 4th to the 8th century, and a humid period from the 12th to the 14th century. These trends correspond to those in other areas of the world.

Chronological details based on pollen analyses in mainland China are scarce, but in Formosa, Tsukada (1966) made pollen analyses and has presented data from 60,000 B.P. He has shown that during Recent times species in the cold temperate zone changed to species in subtropical zones, and again converted to species in the cold temperate zone. These changes occurred gradually.

Climatic changes in Eastern and Central Asia in the postglacial period can be summarized as follows: It was warm in Eastern and Central Asia around 6,000 B.P., and dry in inland Asia at middle latitudes. Then it grew cold in East and Central Asia around 3,000 B.P., and humid in inland Asia at middle latitudes. Inland Asia and West Pakistan were arid around 2,500 B.P. In contrast, it was humid in the Near East, and humid and cold in Japan. In historical times, the warm Little Climatic Optimum in the 8th to 9th centuries and the cold Little Ice Age at the beginning of the 19th century occurred in all parts of Asia.

During this period, however, conditions of precipitation differed from place to place, as shown in Table I.

d. Climatic changes in Southern and Western Asia

According to Brooks (1926), there were three arid periods in Western Asia: the first occurred in 3,700 B.P., the second in 2,100 B.P. and the third in the 3rd century. The latter, especially, extended to Central Asia. South Asia became dry from 600 to 1,200 A.D., as evidenced by the fact that the ruins of Angkor Wat, which was prospering at that time, still exist in the jungle (Willet, 1949).

In West Pakistan around 6,000 B.P., it was humid and the mean annual temperature was probably 2°C or 3°C higher than the present (Lamb et al., 1966). At that time, it was arid in inland Asia, and humid in the Near East. It is thought that in winter, when the Southwest Asian polar frontal zone is stronger, the regions in the Near East and West Pakistan might be humid in contrast to the weak polar frontal zone on the northern side of the Tibetan Plateau resulting in aridity in Central Asia. From about 4,000 B.P., it began to be dry, and this condition increased at about 2,500 B.P. At about the same time it was humid in Japan and the Near East, and arid in inland Asia and West Pakistan. These circumstances may be due to the fact that the climate in about 2,500 B.P. was turning colder. This resulted from an intensification of the anticyclone over the Eurasian Continent in winter and the frequent cold air outflow from the polar region. This accordingly caused an intensification of cyclonic activity along the polar frontal zones over the Mediterranean Sea and East Asia. The humidity therefore increased in the Near East and Japan, while an arid climate took over in Central Asia and West Pakistan.

In the Near East, climatic changes in Recent times have been studied in detail from the standpoint of geomorphology, archeology and history. The period from 11,000 B.P. was divided into eight stages: Remarkable climatic changes occurred from 8,500 B.P. to 4,400 B.P. with rainy weather and higher temperatures; from 4,400 B.P. to 2,800 B.P. there was decreased rainfall; from 2,800 B.P. to 700 A.D. it was humid and warm in winter; from 700 A.D., the rainfall increased a little, while the winters were probably cold from 600 A.D. to 1,000 A.D. (Butzer, 1957, 1958, 1964).

3. PERIGLACIAL LANDFORM IN EAST ASIA

A periglacial landform is formed by repeated freezing and thawing of the surface layer. An active periglacial landform is a useful means for analyzing climate and vegetation in mountains and at high latitudes. In this section, periglacial landforms in Japan and other regions will be examined in terms of the climatic changes.

a. Japan

According to studies by Koaze (1965, 1970) the climatic conditions required for periglacial landform in Japan are such that the soil should be frozen for more than nine months of the year and that there be alternating freezing and thawing for more than three months of the year. The upper limit of the periglacial area corresponds to the snow line on the mountains. On the other hand, the lower limit of the periglacial area corresponds to the upper limit of the forest line or Pinus pumila area. The lower limit is about 2,500–2,700m in the Japanese Alps and 1,600–1,800 m in the Hidaka Mountains of Hokkaido.

Fossil periglacial landforms are an effective indicator of paleoclimate and paleogeography. All fossil periglacial landforms in Japan were formed during or after the last glacial age. The lower limit of the periglacial area in the last glacial age was about 1,800 m lower than the present limit. However, the decrease in temperature in the last glacial age was not equivalent to an 1,800 m height difference. Troll (1944) has published a distribution map of periglacial landforms showing the lower limit of the periglacial area from the Eurasian continent to the African continent, as shown in Fig. 5.

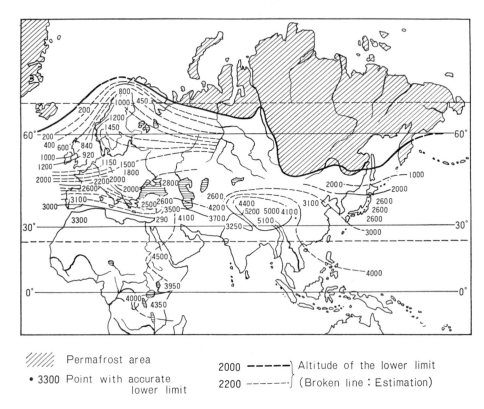

Fig. 5. The periglacial zones of Eurasia (after Troll, 1944).

The question of climatic conditions in the periglacial areas is an important one. Suzuki (1966) has calculated the actual number of freeze-thaw days in Japan. A distribution map showing his results indicates that the maximum is found not at the summit of mountains but on the plateau of eastern Hokkaido. Tabuchi (1970) has also examined climatic conditions in the periglacial areas, and has calculated the number of days when temperatures were at or below the freezing point in Japan. He found that the maximum occurred in the valleys and small basins, because of cold air coming down from the surrounding mountains.

b. *East Asia*

Black (1954) gives a distribution map of the permafrost in the Northern Hemisphere as shown in Fig. 6. The zones of continuous, discontinuous and sporadic permafrost cover about 22% of the area of the Northern Hemisphere. Ray (1951) suggested that permafrost occurred in zones: 1) Continuous permafrost appeared mainly in areas where the climate at present is severe enough to form permafrost. In zone the only unfrozen areas are seen in deep wide lakes and rivers, or in the deep sea. Ground temperatures generally range between $-5°C$ and $-12°C$, rising to $-1°C$ or higher in sporadic regions where mean air temperatures are relatively higher.   2)   In the second zone discontinuous permafrost appears in small, scattered, unfrozen areas.   3)   In the third zone sporadic or small islands of permafrost occur generally in unfrozen areas. Many of these islands appear to be relics of a preceding colder climate and steadily decrease in size. The distribution of permafrost in the USSR is sufficiently well known that the isopleths of thickness can be drawn (Brown, 1960).

In East Asia, permafrost extends far south of the $-2°C$ isotherm. It is thought that the

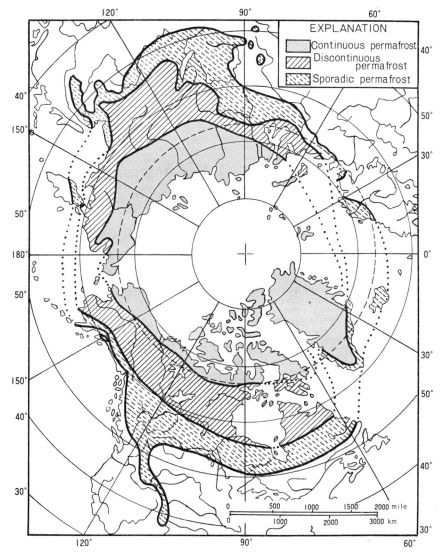

Fig. 6. Contemporary extent of permafrost in the Northern Hemisphere (after Black, 1954).

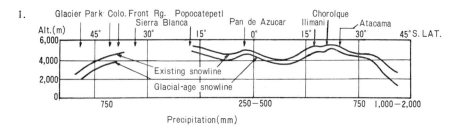

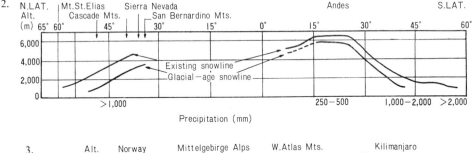

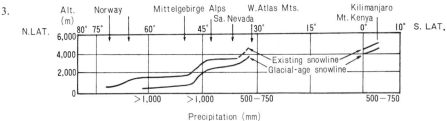

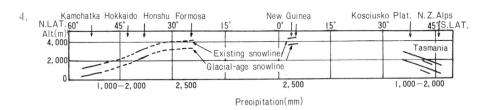

Fig. 7. Curves showing existing the climatic snow line, and the glacial age snow line inferred from the altitude of cirque floors, in four north-south belts extending through both polar hemisphere (after Flint, 1963). 1. Continental Americas, 2. Maritime Western Americas, 3. Maritime Western Europe; East Africa, and 4. Maritime Eastern Asia; Australasia.

climatic conditions leading to permafrost in this region are not in equilibrium at present, but that the permafrost is a relic of low temperatures in the Pleistocene. A number of workers have also tried to study the correlation of permafrost occurrence with other climatic variables, especially indices of freezing or thawing. No definite relationships have been established from these analyses. On the other hand, the extensive thickness and location of some southerly permafrost areas in the USSR have led to the view that the permafrost in these areas is essentially old, and that it is now slowly degrading with the amelioration of climate since the last glacial period.

The results of studies in the USSR suggest that there is an important contrast between the southerly zone of degrading permafrost and the northerly zone of active or aggrading permafrost. These two zones are separated by an intermediate belt in which permafrost is being neither actively formed nor actively destroyed at the present time.

4. GLACIAL AGE CLIMATE IN ASIA

The climate of the glacial age may also have been influenced by the movement of frontal zones. The mechanism of atmospheric circulation in the glacial age was probably the same as at present. Suzuki (1970), referring to Markov's report, pointed out that the polar frontal zone, in the glacial age extended over Kyushu. The southern limit of the arctic frontal zone nearly coincides with the southern limit of the Tundra area and did not cross the Himalayas in the glacial age.

Flint (1963) illustrates the snow lines over the continents in the present and the glacial age. The snow line in maritime East Asia is lower than in continental and maritime western

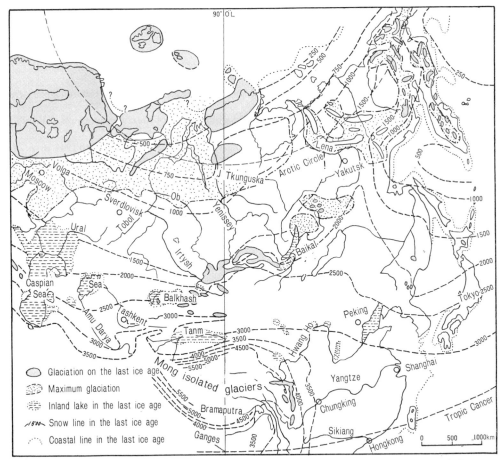

Fig. 8. Glaciers, snow lines, inland lakes and others in Siberia, Central and East Asia (after Frenzel, 1960). Snow line is given in m and coastal line in the last ice age shows 100 m isodepth.

America, but it is the same as that in maritime western Europe, as shown in Fig. 7. Frenzel (1960) gives the distribution of the glaciers, snow lines and inland lakes in the glacial age in Siberia, Central Asia, and East Asia. Glaciers formed during the last glacial age or the maximum glacial age are shown in Fig. 8. The distribution of the glaciers in Asia is inaccurate as compared with those in Europe. The position of the inland lakes in the maximum glacial age is accurate for West Asia and Europe, but incorrect for East Asia. The snow line for the last glacial age in East Asia is accurate, however, the snow line in the Japanese Alps was about 2,500 m and the snow line in the Hidaka Mountaines was about 1,600 m. The coastal line in the last glacial age is also illustrated in the figure.

ACKNOWLEDGEMENT

The authors wish to thank the numerous persons who have assisted in various ways during the preparation of this paper, and especially Professor Fumio Tada, Komazawa University, and Professor Masatoshi M. Yoshino, Hosei University, for their continuing guidance and encouragement. Dr. Iwao Tsuchiya, Meteorological Research Institute, Tokyo, also provided valuable assistance. For the drawing of the figures, the authors are indebted to Mr. Katsuhisa Fukuda.

References

Arakawa, H. 1954: Fujiwara on five centuries of freezing dates of Lake Suwa in the Central Japan. *Arch. Met. Geophy. Biokl.* (B) **6**(1) 152–166.

Arakawa, H. 1955: Climatic change as revealed by the blooming dates of the cherry blossoms at Kyoto. *J. Geog.* **64**(2) 31–32.

Arakawa, H. 1956: On the secular variation of annual totals of rainfall at Soul from 1770 to 1944. *Arch. Met. Geohy. Biokl.* (B) **7**(2) 205–211.

Arakawa, H. 1960: Japanese tree-ring analyses. *Arch. Met. Geophy. Biokl.* (B). **10**(2) 210–212.

Asakura, T. 1971: Distribution of cloudiness and precipitable water during Baiu season and their variation over Monsoon Asia. (In this volume)

Black, R. F. 1954: Permafrost—A review. *Geol. Soc. Amer., Bull.* **65** 839–856.

Brooks, C. E. P. 1926: Climate through the ages (London).

Brooks, C. E. P. 1951: Geological and historical aspects of climatic change. *Compend. Meteorology, Amer. Meteorol. Soc.* 1004–1018.

Brown, R. J. E. 1960: The distribution of permafrost and its relation to air temperature in Canada and the U.S.S.R. *Arctic* **13** 163–177.

Butzer, K. W. 1957: Late glacial and Postglacial climatic variation in the Near East. *Erdkunde.* **11** 21–35.

Butzer, K. W. 1958: Quaternary stratigraphy and climate in the Near East. *Bonner Geogr. Abh.* (24) 157p.

Butzer, K. W. 1964: Environment and archeology—An introduction to Pleistocene geography. Aldine, Chicago 524p.

Chappell, J. E., jr. 1970: Climatic change reconsidered: Another look at "The Pulse of Asia." *Geogr. Rev.* **60** 347–373.

Chu, Co-Ching 1954: Climatic changes during historic time in China. *In* Collected Sci. Papers, Meteorology 1919–1949, 265–279.

Flint, R.F. 1963: Glacial and Pleistocene Geology. 3rd Ed. (John Wiley & Son. Inc.) 553p.

Frenzel, B., Troll, C. 1952: Die Vegetationen des nordichen Eurasines während der letzten Eiszeit. *Eiszeitalter Gegenwart* **2** 154–167.

Fuji, N. 1965: Palynological studies on the late Tertiary and Quaternary systems in Hokuriku district, Central Japan. *Quat. Res.* **4**(3–4) 183–190.

Fujii, N. 1966: Climatic change of Postglacial age in Japan. *Quat. Res.* **5**(4) 149–156.

Horie, S. 1964: Kosho hattatsushi kenkyujô no nendaishiryo (Chronological data in a historical study of the development of lakes). *Kagaku* **34**(2) 98.

Horie, S. 1965: Late Pleistocence glacial fluctuations and changes of sea level in the Japanese Islands and their tentative correlation with oscillations in North America and Europe. 6th Internat. Quaternary Cong., (Warsaw 1961), Rept., Vol. 1, Commission on the Quaternary Shorelines, 175–184.

Horie, S. 1967: Late Pleistocene climatic change inferred from the stratigraphic sequence of the Japanese lake sediments: Means of correlation of Quaternary successions, INQUA 7th Congr. Proc., Vol. 8, Salt Lake City, University of Utah Press.

Horie, S. 1968: Nihon no neoglaciation ni tsuite (On the Neoglaciation in Japan). *Kyoto Univ. Bôsaiken-kyusho Nenpo* **11**(B) 47–58.

Hoyanagi, M. 1961: Kikôhenka eno kanshin (Concern for climatic changes). *Chiri* **6**(7) 7–14.

Hoyanagi, M. 1965: Sand-buried ruins and shrinkage of river along the old Silk Road region in the Tarim basin. *J. Geogr.* **74**(745) 1–75.

Hoyanagi, M. 1966: Glacial problems in China. (1), (2) *J. Geogr.* **75**(752, 753) 155–165, 210–225.

Kaizuka, S. 1964: Tokyo no shizenshi (Natural history of Tokyo). Kinokuniya 186p.

Kaizuka, S. 1969: Henka suru chikei (Geomophologyical changes). *Kagaku* **39**(1) 11–19.

Kaizuka, S. 1970: Nihon no Würm hyoki no kikôchikeiku (Classification of climatic-geomorphology in Japan during the Würm glacial age. *Geogr. Rev. Japan* **43**(2) 116–117.

Koaze, T. 1965: The patterned grounds of the Daisetsu Volcanic Group, Central Hokkaido. *Geogr. Rv. Japan* **38**(12) 179–199.

Koaze, T. 1970: Nippon no syûhyôgachikei to sono keiseijyôken. *Geogr. Rev. Japan* **43**(2) 107–109.

Kobayashi, K., Hoshiai, S. 1955: Hyoki, kanpyoki oyobi kôhyoki no kikôhenka (Climatic changes in the glacial, interglacial and postglacial ages). Nihon gakujutsukaigi, Daiyonbu, Symposium 25–34.

Kobayashi, K. 1962: Dai Yonki (The Quaternary period) (Vol. 1). Chigakudantaikenkyukai 194p.

Lamb, H. H., Lewis, B. P. W., Woodroffe, A. 1966: Atmospheric circulation and the main climatic variable between 8000 and 0 B. C. *In* Meteorological evidence, World Climate from 8000 to 0 B. C. Roy. Meteorol. Soc. 174–217.

Matusima, S. 1941: Betrachtung zur Waldentwicklung in Korea und Grund von Pollenstatistik. *J. Jap. For. Soc.* **28**(8) 441–450.

Maejima, I 1966: Some remarks on the climatic conditions of Kyoto during the period from 1474 to 1533 A.D. *Geogr. Rep. Tokyo Metropolitan Univ.* (1) 103–111.

Nakamura, J. 1952: A comparative study of Japanese pollen record. *Res. Rep. Kochi. Univ.* **1**(8) 1–20.

Nakamura, J. 1967: Kafun Bunseki (Pollen analysis). Kokinsyoin 232p.

Nishimura, K. 1965: Climato-genetic geomorphology in Japan. *Sci. Reps. Tohoku Univ. 7th Ser.* (Geogr.) 14.

Nishioka, H. 1949: Kandan no rekishi (History of warm and cold days). Kogakusya 248p.

Ray, L. L. 1951: Permafros. *Arctic* **4** 196–203.

Sakaguchi, Y. 1963: On postglacial sea level changes in Japan. *Quat. Res.* **2**(6) 211–219.

Schmitthenner, H. 1927: Die Oberflächengestaltung im aussertropischen Monsunklima. Düsseldorfer Geographische Vorträge und Erörterungen. 26–36.

Schmitthenner, H. 1930: Der Weitaishan. Ein Reise auf der heiliger Berg des Wides in Nord China. *Mitt. Ges. f. Erdk. z. Leipzig* 50.

Schmitthenner, H. 1932: Landformen in aussertropischen Monsunsgebiet. Beobachtung und Untersuchung in China. *Wiss. Veröff. Mus. f. Länderkunde z. Leipzig.* N. F. 1.

Suzuki, H. 1960: Periglaziale Erscheinungen in Nord-Hokkaido. *Geogr. Rev. Japan* **33**(12) 625–628.

Suzuki, H. 1962: Southern limit of peri-glacial landform at low level and the climatic classification. *Geogr. Rev. Japan* **35**(2) 67–76.

Suzuki, H. 1970: Daiyonki kikotaihendo ni kansuru chiken to Nippon no ichi. *Geogr. Rev. Japan* **43**(2) 117–118.

Suzuki, H. 1966: Distribution of freeze-thaw days in Japan. *Geogr. Rev. Japan* **39**(4) 267–270.

Tada, F. 1938: Über die periodische Änderung der Regenmenge in Chosen seit dem Jahre 1776. Comptes Rendus du C. R. Congr. Inter. Geogr. Amsterdam 1938. Bd. 11 Trauvaux Sections A-F. Leiden 305–308.

Tada, F. 1950: Geomorphology of the northern Manchuria plain. *Bulletin of the Geogr. Inst., Tokyo Univ.* (1) 133–145.

Tada, F. 1964: Shizenkankyô no henbô (Changes in the natural environment). *University of Tokyo Press* 282p.

Tabuchi, H. 1970: The distribution of degree days of freezing in northeast Japan. *Climatological Notes, Hosei Univ.* (5) 30–32.

Taguchi, T. 1939, 1939, 1940, 1940, 1940, 1941, 1942: Nihon no rekishijidai no kikô ni tsuite (1)–(7). (On the climate in Japan in historical time) *kaiyokisyodai iho* (124) 1–30, (126) 1–11, (130) 1–8, (132) 1–12, (133) 1–15, (137) 1–6, (140) 1–11.

Troll, C. 1944: Strukturböden, Solifluktion und Frostklimate der Erde. *Geol. Rundschau* **34** 545–694.

Troll, C. 1948: Der subnivale oder periglaziale Zyklus der Denudation. *Erdkunde* **2** 1–21.

Tsukada, M. 1958: On the climatic change of postglacial age in Japan, based on four pollen analyses. *Quat. Res.* **1**(2) 48–58.

Tsukada, M. 1966: Late pleistocene vegetation and climate in Taiwan (Formosa). *Proc. Nat. Acad. Soc.* **55**(3) 543–548.

Willet, H. C. 1949: Long-period fluctuations of the general circulations of the atmosphere. *J. Met.* **6** 34–50.

Wissmann, H. v. 1938: Über Lössbildung und Würmeiszeit in China. *Geogr. Zeitsshr.* **44** 201–220.

Wissmann, H. v. 1959: Die heutige Vergletscherung und Schneegrenze in Hochasien mit Hinweisen auf die Vergletscherung der letzten Eiszeit. *Akad. Wiss. u. Lit. Mainz Abh. math-nat. Kl.* (14) 1–307.

Wissmann, H. v. 1960, 1961: Stufen und Gürtel der Vegetation und des Klimas in Hochasien und seinen Randgebieten. *Erdkunde* **14** (4) 247–272, **15** (1) 19–44.

Yamamoto, T. 1967: Some considerations on the long-term variation of rainfalls in the historical time of Japan and its surroundings. *Quat. Res.* **6**(2) 63–68.

Yamazaki, T. 1940: Beiträge zur Verwandlung der Baumarten im südlichen Teile von Korea durch die Pollenanalysis. *J. Jap. For. Soc.* **22**(2) 73–85.

PART V

REPRESENTATION OF WETNESS AND DRYNESS OF
MONSOON ASIA BY CLIMATOLOGICAL INDICES

REGIONAL DIVISIONS OF MONSOON ASIA
BY KÖPPEN'S CLASSIFICATION OF CLIMATE

Mitsuharu MIZUKOSHI

Abstract: In this paper, the writer gives detailed maps of climatic divisions, revising Köppen's classification of climates in Asia. A revised map of climatic divisions was drawn using all available data, and a comparison was made with previously published maps. A map of climatic divisions according to the "year climatic method" was also devised, and compared with the map based upon the original method of Köppen's classification with normal climatic values. Some differences between the two maps of climatic divisions can be pointed out, especially the climatic types that have marked seasonal variation in precipitation tend to occupy wider areas in climatic divisions by the year climate method. Lastly, the appearance frequencies of each climatic type in the year climate were calculated at various stations, and on this basis the core regions of each climatic type were ascertained.

1. FOREWORD

Among the various types of classifications of climate, Köppen's classification is the most widely used for world maps of climatic divisions. Köppen's classification of world climates was first published in 1918. A number of revisions have been made since then. For example, the boundaries between the dry or B climates and the moist tree climates or between the warm or C climates and the cold or D climates have been altered. But the principles of this classification are still widely accepted, for the following reasons: 1) The classification is based upon annual and monthly means of temperature and precipitation. Therefore, the processes of classification are relatively easy to apply. 2) Of the several climatic elements, temperature and precipitation are the most significant, and data of this sort are available for a great number of regions. Therefore, detailed maps of climatic divisions based on these parameters can easily be made. 3) Each climate can be objectively assigned to a different type. To define the boundaries of climatic types, critical values of temperature and precipitation are employed. 4) The climatic boundaries agree well with other boundaries of natural environment such as vegetation and soil.

There are some demerits to the classification, however. The main objections are as follows: 1) Theoretical conditions relating to the formation of climates are nelgected. For instance, high altitude climate is treated in the same way as high latitude climate. 2) Some disagreement exists between climatic divisions and features of the natural and cultural landscape, especially in the southern hemisphere. 3) The deficiency of climatological data for large parts of the world makes a climatic classification with rigid boundary criteria unsatisfactory.

Some revised methods have been published: for instance, the concept of the year climate. As stated above, the deficit of available data for large parts of the world makes it impossible to devise completely accurate climatic divisions. Data for annual and monthly means of temperature and precipitation are not sufficiently abundant to draw detailed maps for large parts of Asia, although this can be done for limited areas, such as Japan, India, eastern China and Indonesia.

2. REVIEWS OF PAST STUDIES ON CLIMATIC DIVISIONS OF MONSOON ASIA AND THE PURPOSE OF THE PRESENT STUDY

When Köppen's classification was published in 1918, climatic regions were designated for the entire world; but for Asia, the designation was very rough, because of the scarcity of available data. For example, the Cw climates covered large parts of China and the Cf climates were located on the coastal areas of central China, whereas the Cf climatic regions expand into the interior of China in present maps. The classification was subsequently revised and extended by Köppen and his successors to become the most widely used of climatic classifications for geographical purposes (Geiger and Pohl, 1954).

Finch and Trewartha (1936) revised Köppen's classification, omitting the Cw climates on the grounds that they were not sufficiently distinctive to warrant setting them apart as separate types. This left large parts of East Asia belonging to one climatic region (humid subtropical climate), and large parts of the Indian Peninsula belonging to the tropical savanna. Compared to present maps, regions of undifferentiated highlands were very wide and tropical regions were shown to be homogeneous.

Recently, a newly revised world map according to Köppen's classification was published (Köppen and Geiger, 1961), using new data. Precise boundaries were drawn taking altitude into consideration, but in the tropical area, the divisions were still not very precise.

The present writer has attempted to make more detailed maps of climatic divisions and to revise Köppen's classification of climates in Asia in this paper. For Monsoon Asia, the range from 10°S to 50°N and from 65°E to 175°W was taken. There are a number of stations where temperature and precipitation data are available at present. But the distribution of stations is still very uneven. About 600 stations were selected, taking into consideration map scale, density and the distribution of stations. Temperature and precipitation data from these stations were used to draw a detailed new map for Monsoon Asia according to Köppen's system as given in the next section. The map clarifies schema of water supply and demand in these areas. An attempt has also been made to revise the climatic division.

3. KOPPEN'S CLASSIFICATION OF CLIMATES IN MONSOON ASIA

Areas with a high density of climatological observations in Monsoon Asia are Japan, Korea, Northeast and East China, Java, the Philippines, and India. Areas where the density of observation points is very low are the highlands and arid areas in Central Asia, including the Tibetan Plateau, the Himalayas, the Mongolian Plateau, and the Tarim basin, and the tropical islands in Southeast Asia such as Sumatra, Borneo, and New Guinea.

The revised map is shown in Fig. 1. Comparing this map with former maps, the following differences can be pointed out:

(1) Boundaries between the dry (B) climates and the humid (A, C, and D) climates have been clarified, and the boundaries between the B (or BS) and C climates in particular have

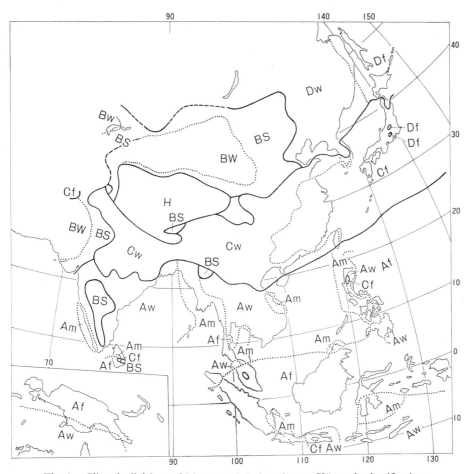

Fig. 1. Climatic divisions of Monsoon Asia based upon Köppen's classification.

been made more rigid. The dry climates in the middle latitudes are found usually in the deep interiors of continents, for instance in Central Asia. In previous maps of climatic divisions, the steppe (or BS) climate is restricted to areas far from Pohai Bai (at a distance of about 500 km), in northern China. Geiger's revised map shows that the BS climate extends to the northern and southern coasts of Pohai Bay, and the D climates are located on the west coast. In Fig. 1, the climatic distribution pattern is similar to Geiger's map, but the extension of the BS climate to the northern coast of Pohai Bay has been eliminated. Furthermore, the eastern limits of the middle latitude BS climate extend close to the coast of Pohai Bay; in particular, the BS climate occupies the vicinity of the Shantung Peninsula and the mouth of the Hwan River.

The regions of arid or desert climate (BW climate) that are found in the interior of Asia extend eastward. This fact was already shown in Gieger's revised map (1961). Since the available climatic data are very scanty, the boundaries between the BS climate and the BW climate are uncertain.

With respect to the regions of tropical and subtropical dry climates, there are no marked differences from previous maps. There is, however, a small region of BS climate in the vicinity of Mandalay, Burma, and it may be an interesting question for further study why a small pocket of dry climate should appear in the area.

(2)   The regional distributions of the tropical (A) climates are more complex. The tropical climates are divided into three types: 1)   Tropical rainforest (Af) climate. The Af climate appears mainly near the equator within the equatorial rain belt. Typically, there is heavy precipitation (about 2,000 mm or more) throughout the year, with no dry season. The main regions of this type are Ceylon, Sumatra, Borneo, New Guinea, and the Philippines.   2) Tropical monsoon (Am) climate. This type has seasonal heavy rainfall and a short dry season. Representative regions are tropical coasts backed by highlands and facing away from the Pacific: for instance, the west coasts of India and Indochina. In these areas, onshore monsoon winds bring a strongly seasonal precipitation.   3)   Tropical savanna (Aw) climate. This type of climate differs in two principal respects from the Af type. (a) There is usually less precipitation, and (b) rainfall is unevenly distributed throughout the year, i.e. there is a distinctly wet and a distinctly dry season. Tropical regions with Aw climate are the northern parts of the west coast of India, the east coast of India, Burma, Thailand, Laos, Cambodia, Vietnam, and the eastern part of Indonesia.

It has been supposed that the distribution of these climatic regions in Monsoon Asia is quite complex. The climatic boundaries are not clear, because of insufficient data. The writer has reexamined these boundaries, however, in the light of new materials. As a result, it has been shown that in the Philippines, for example, the coastal regions of the Pacific Ocean belong to the Af climatic type, and the coastal regions of the South China Sea belong to the Am climate. Between these regions there are narrow zones of the Aw climatic type. A detailed map of climatic divisions on Java was presented earlier by Fukui (1956). Fig. 1 of the present paper shows the same features as his map, locating the three types of A climate on the island. In other areas of Indonesia and its vicinity—Sumatra, Borneo, Celebes, and New Guinea for example—the climatic boundaries are uncertain. Much data is still required before any definite boundaries can be assigned.

In India and Indochina, the patterns for the A climatic types given in Fig. 1 do not differ significantly from the maps published previously. But in India, the boundaries between the C and A climates are shifted to the A climates, and the regions of the C climates extend to the vicinity of the coast of the Bay of Bengal.

(3)   The warm temperate rainy climates (the C climates) are divided into three types, according to seasonal distribution of precipitation: 1)   The Cf, or warm temperate rainy climates without any dry season;   2)   the Cw, or warm temperate rainy climates with dry winters;   3)   the Cs, or warm temperate rainy climates with dry summers.

Of the three types, the Cf and Cw climates occur in Monsoon Asia. The Cw climates are mainly found in Northern India and greater China, and all other humid areas of the temperate zone belong to the Cf climates. In the early maps, the Cw climates covered all of central and southern China. But as more data have been accumulated for these areas the Cf climatic regions have been expanded into greater China. According to Geiger's revised map, the Cf climates cover a large part of central and southern China, excluding the coastal regions of the South China Sea.

Figure 1 differs in other respects from Geiger's revised map. On the northern side of the Yangtze River, the Cf climates cover smaller areas. In other regions, for instance Japan and its vicinity and northern India, there are no major differences between the two maps.

The boundaries between the C climates and the B and D climates are somewhat uncertain, especially in the interior of China. The Tibetan Plateau and its surroundings belong to the polar (EH) climate due to the elevation of the land. The boundaries between the EH and C climates are also uncertain. In previous maps, the EH climatic regions bordered immediately onto the regions of the C climates. But between these two regions, B or D climates may also exist. Lhasa and Gyantze (southern Tibet) belong to the BS climates, and Taining and Omeishan (interior China) belong to the Dw climates. Although the regions of B or D climates may be very narrow, their existence can not be neglected. With respect to the snow-forest or D climates, no differences were found from earlier work.

4. CLIMATIC DIVISIONS OF MONSOON ASIA BY THE YEAR CLIMATE METHOD

Köppen's classification of climates is based upon normal values of annual and monthly temperatures and precipitation. The normal values are calculated as the mean values for several years. If Köppen's classification is used to assess the climate for one year alone, the climatic type of that year can be determined. This is called "the year climate." Although the year climate varies from year to year, it is possible to express the climatic type of a station by the type of year climate which predominates. Although the prevailing type of year climate does not always agree with the normal climatic type, it does represent the characteristics of climate at each station. The concept of year climate was first proposed by Russell (1932). He described the dry climates of the United States, by the year climate method. Sekiguchi (1951) worked with the year climate method in Japan. He investigated the distribution patterns of the year climate type for each year, and observed frequencies of various climatic types at each station. He also divided Japan into several climatic regions, using the appearance frequencies of the year climate at each station, and pointed out that ordinary climatic types under average conditions were Cfa (or Dfa) at almost all stations in Japan. But he found many years with Cw or Cs climates.

Using the year climate method, the writer has attempted to represent the climatic type of each station, and to divide Monsoon Asia into several climatic regions. To determine the prevailing type of year climate, annul, and monthly data of temperature and precipitation for many years are necessary. The writer collected annual data for these climatic factors for the period 1931–1960 in the areas investigated. The main data sources are listed as shown later. It was difficult to collect complete data at every station, because some areas lack the climatological data for the period. Therefore, only those stations at which data for five or more years could be gathered were chosen. The number of stations was about 450. The climatic divisions for Monsoon Asia based on the prevailing type of year climate are shown in Fig. 2. Certain features of the map differ from the map in Fig. 1, which was drawn according to normal methods. The main differences between the two types of climatic division are as follows:

a. *Expansion of the Regions of Cw Climates*

The Cw climates have a large seasonal variation of precipitation, that is, a dry winter and wet summer. The Cw climates are seen mainly in northern India and southern China on the map of climatic types based upon normal values (Fig. 1), and were therefore called the "China Climate." However, large parts of eastern China are covered by the Cf climates as given in Fig. 1. The Cw climates appear only in the area along the lower course of the Hwan River, the interior of China and the coastal areas of the South China Sea. Fig. 2 shows, however, that the regions of Cw climates are widespread throughout East Asia. In addition, the following four points should be noted: a) The Cf climatic region of China in Fig. 1 is

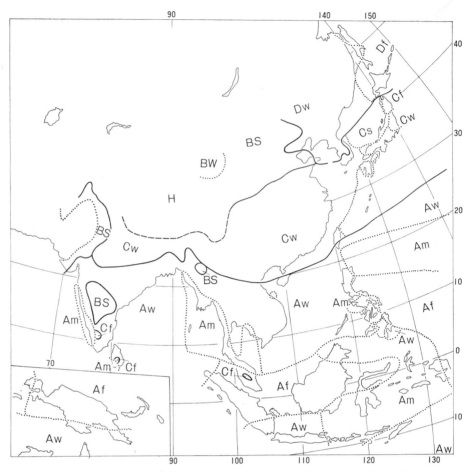

Fig. 2. Climatic divisions of Monsoon Asia based upon the prevailing type of the year climate.

replaced completely by the Cw climatic region in Fig. 2. In other words the climates in central and southern China are characteristically Cw climates in each year. b) In southern Korea, the Cf climatic region also changes to Cw. c) In Formosa, the Cf climatic region disappears in Fig. 2, except in the east coastal area. d) In Japan, the distribution of climatic regions becomes more complex.

Accroding to the normal representation by Köppen's classification, the Japanese Islands belong to the Cf climates, except for Hokkaido, a small region of northeastern Honshū and the highland areas of Honshū as given in Fig. 1. But using the prevailing year climate as given in Fig. 2, the Cw climatic regions are more widespread in Japan. The main regions of Cw climates are the greater part of Kyūshū, part of Shikoku, the western part of Chūgoku, the northern part of the Kantō and the southern part of the Tōhoku districts. The regions of Cf climates are restricted to the central and northern parts of Japan. It is interesting that the Cs climates appear on the Japan Sea side, but this feature will be treated in a later section. The distributions of Cw climatic regions in Japan have already been treated by Sekiguchi (1951), who presented nearly the same results as the present ones. In the southern part of the Tōhoku and Tōkai districts, however, the results are somewhat different.

b. *Expansion of the Regions of Aw Climate*
In Fig. 1, the regions of Aw climate appear mainly in the southeastern part of the Indian Peninsula, the greater part of Indochina the southwestern part of the Philippines, the Little Sunda Islands and the southern part of New Guinea. In Fig. 2, however, the regions of Aw climate are more extensive: for instance, the northern coast of the Bay of Bengal, the east coast of Indochina, the South China Sea side of the Philippines and Java. The regions of Am climate in Fig. 1 are replaced by regions of Aw climate. Generally, the regions of Am climate become narrower in Fig. 2.

c. *Appearance of the Cs Climates in East Asia*
On the map of prevailing year climates given in Fig. 2, a region of Cs climates appears in the southern part of the Japan Sea. In the distributions following Köppen's classification, the Cs climates appear mainly on the west coast of the continents. The Cs climates occur in the Mediterranean region including Western Asia, and on the west coast of the United States. The Cs climates do not appear in the eastern part of the continents. However, the characteristics of the Cs climates that appear in Fig. 2 may differ from the original Cs climates, the Mediterranean climates, for the following reasons: 1) The total amount of annual precipitation is much larger in Japan (1,500–3,000 mm) than in the Mediterranean (500–1,000mm).   2)   Most of the total precipitation in Japan is snowfall in winter.   3)   A considerable amount of summer precipitation is also expected.   4)   The distribution of Cs climates is closely related to the seasonal shifting of the subtropical anticyclones. The area covered by the subtropical anticyclones or located in their peripheries in summer have the Cs climatic type. Summers are very dry and fine because of the anticyclones and winters are rainy because of the extratropical cyclones. In contrast, the climates of the areas along the Japan Sea coast have no direct connection to the subtropical anticyclones, but the north-westerly winter monsoon together with the great amount of moisture suplied from the Japan Sea surface causes heavy snowfalls in winter. Hence it can be said that the Cs climates appearing on the Japan Sea coast differ from the original Cs climates, although Köppen's method of classification cannot distinguish between them.

d. *Contraction of the Af and Cf Climatic Regions*
The contraction of the Cf climatic areas was discussed in the preceding section. The Af climatic areas become narrower and Af climate is replaced by Am climate in the southern Philippines, the Moluccas and the northern Mariana Islands. In Ceylon, the Af Climatic disappears.
In general, the climatic types that have marked seasonal variation of precipitation tend to occupy wider areas in climatic division by the year climate method. This is probably because if the maximum or minimum precipitation amounts appear in different months in each year, the differences of monthly precipitation are smoothed out when the normal values are calculated. Therefore climates of the "f" type become in conspicuous. If the prevailing year climate at a station is of the "s" or "w" type, it retains its type. Climates of the "s" or "w" types, therefore, become more extensively distributed in climatic division based upon the prevailing year climates.

5. REVISED CLIMATIC DIVISIONS OF MONSOON ASIA BY THE YEAR CLIMATE METHODS

The method of climatic division based upon the prevailing year climate, however, has some demerits. If the frequency of appearance of the most prevalent type of year climate is far above that of the next prevalent type, the most prevalent type probably represents accurately the climate at the station. But if the difference in frequency of appearance between the most prevalent type and the next is relatively small, it is difficult to determine the representative climatic type. Accordingly, when the frequency of appearance of the most prevalent climatic type differs from the next most prevalent by more than 30%, the most prevalent type is defined as the representative climatic type at the station. But when the difference is less than 30% the station is considered to have a transitional climatic type. Two or three types of climate are described to represent the climate at the station. Figure 3 indicates the revised climatic divisions of Monsoon Asia based on this method.

Regions with a transitional climate are found in many areas, as seen in Fig. 3. The most complex of these are Japan and its surrounding regions, where the Cf, Cw, Cs, Df, Ds, and Dw climates all appear. This indicates that the year climate varies greatly from year to year in these regions. In some areas, the transitional zones are imperfect, probably due to insufficient data.

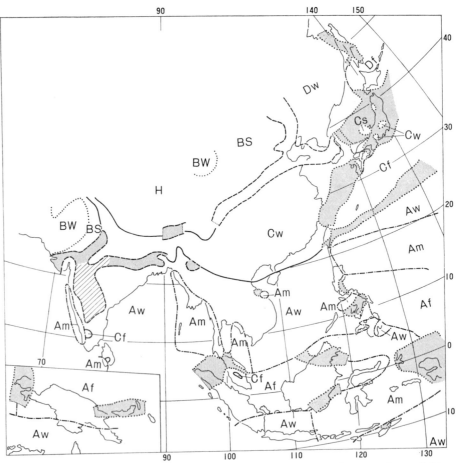

Fig. 3. Revised climatic divisions of Monsoon Asia based upon the year climate (with reference to the appearance frequencies of the most prevalent and next prevalent types of year climate).

6. The Distribution of Multiple Occurrence of Climatic Types in the Year Climate

If many types of year climate appear at a station from year to year, the station is considered to have an unstable climatic condition. The number of climatic types that appear in the year climate show the stability of the climate. Figure 4 is a distribution map showing regions in which more than three types occur. As climatic types, the following were chosen: Af, Am, Aw, BS, BW, Cf, Cw, Cs, Df, Dw and Ds.

Areas located in the vicinity of major climatic boundaries, for example, between A and C climates, and between C and D climates, have many types of year climate. Within the areas of A or C climates, there are also some areas that have many types of year climate. In Japan and its vicinity, many stations have the three types of C climate, Cf, Cw, and Cs. In part of the Philippines, the area from northern Sumatra to northern Malaya, southern Borneo, northern Celebes, and northeastern New Guinea, many stations have the three types of A climate, Af, Am, and Aw. These regions have an unstable climate.

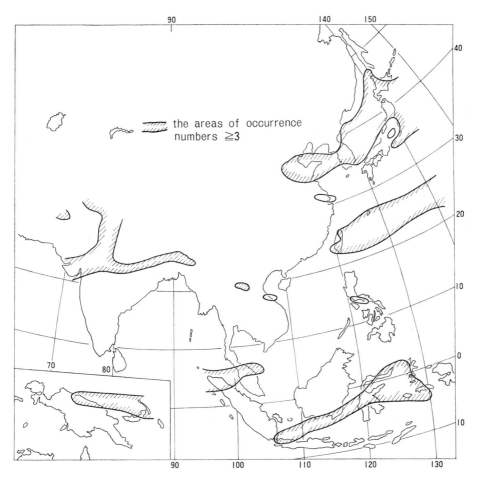

Fig. 4. Areas where more than three climatic types appear in the year climate.

7. THE DISTRIBUTION OF APPEARANCE FREQUENCY OF EACH CLIMATIC TYPE IN THE YEAR
    CLIMATE

It was shown earlier that stable regions have only one or two climatic types of year climate
from year to year. In this section, the writer has attempted to distinguish the core regions of
each climatic type. For this purpose, distribution maps of appearance frequencies were
drawn for each climatic type.

a. *The A Climates* (Fig. 5)
The A climates appear widely in the low latitudes. Their northern limits are about 25°–
28°N in the western Pacific Ocean. To the west, the limits shift to the south. The limits are
about 20°N in Southeast Asia. In India, the limits shift to the north again, locating at about
25°N.
The zones in which A, B or C climates appear in the year climate of each year are defined
as changeable climatic zones. The width of the zone, that is, the distance from 0% appearance of A climates to 100%, is relatively narrow in Southeast Asia and the Pacific Ocean.

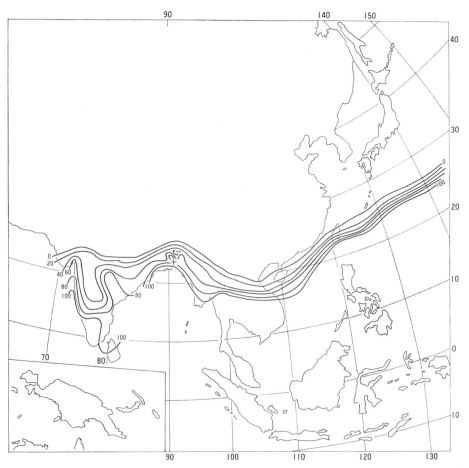

Fig. 5. Appearance frequencies of the A climates in the year climate.

The width is about 300–500 km. In western Indochina and India, the width becomes greater. In India, the width extends to 2,000 km. In the greater part of India, A and B climates appear in the year climate. From this it can be concluded that the Indian climate is unstable because of the variable amount of precipitation.

b. *The Aw Climate* (Fig. 6).

The Aw climate is constant from year to year in Indochina and on the coast of the Bay of Bengal. In the tropics the Aw climate is prevalent, except for the equatorial zones where Af climates prevail. The zones with maximum frequencies of appearance of Aw climate are in the subtropical areas of the Northern and Southern Hemispheres. In the Northern Hemisphere, India, Thailand, Laos, Cambodia, Vietnam, the Philippines and the islands up to about 25°N are predominantly areas of Aw climate. In the Southern Hemisphere, although the main area is off the map in the figure, the region from Indonesia to the southern coast of New Guinea has prevailing Aw climate. Between these two main zones, a less marked zone extends from the Philippines to New Britain. The zonal structure of the Aw climatic distribution is disturbed in the Northern Hemisphere.

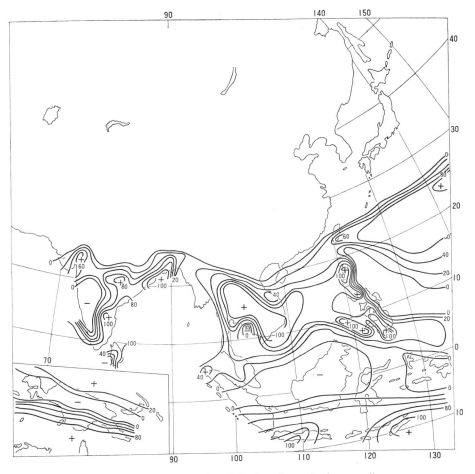

Fig. 6. Appearance frequencies of the Aw climate in the year climate.

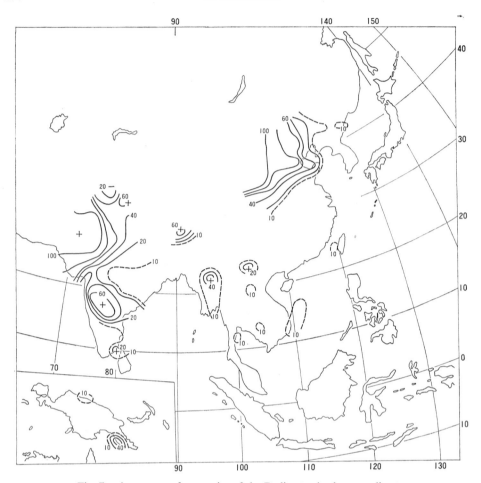

Fig. 7. Appearance frequencies of the B climates in the year climate.

c. *The B Climates* (Fig. 7)

The B climate areas spread primarily over Central and West Asia. In India, especially, the B climate areas are widely distributed. The precipitation is relatively unstable in the area. In other regions, the climatic areas are scattered, for example in Burma, parts of Indonesia, New Guinea and even the islands in the Pacific Ocean. The reason why the B climate areas are located in the center of the Pacific Ocean is ascribed to the position of the subtropical anticyclones over the Northwest Pacific Ocean.

d. *The Cw Climates* (Fig. 8)

The Cw climates appear in China and on the southern side of the Himalayas. They are known as the "China Climate," as mentioned earlier. But in normal maps based on Köppen's climatic divisions, the Cw climates are not extensive enough to ocver the whole of China, while in the map of climatic division according to the year climate, the Cw climates spread over all of China. Moreover, the Cw climates appear widely in the temperate zone, except for the Japan Sea Coast of Japan, so that they are the most common type in temperate Asia. The Japan Sea coast areas of Japan, however, have very different climatic features.

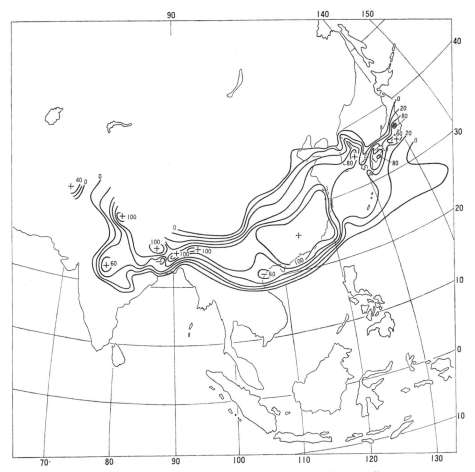

Fig. 8. Appearance frequencies of the Cw climates in the year climate.

8. SUMMARY

Climatic divisions of Monsoon Asia were determined by several methods. The main con-conclusions can be summarized as follows:

1) New climatic divisions using Köppen's method with normal values of temperature and precipitation revealed some differences from earlier ones. a) The boundaries between the B and C climates were shifted towards the C climatic regions. b) The boundaries between the BW and BS climates were shifted towards the BW climatic regions. c) In subtropical areas, isolated BS climatic areas exist. d) Detailed examination of the A climate showed that the regions of Af, Am, and Aw climates are complex in their occurrence in tropical regions.

2) Climatic divisions based upon the prevailing year climate are very different from divisions based upon normal values using the original Köppen method. a) In general, the "w" or "s" types of climate cover greater areas, while the "f" types of climate cover smaller areas than previously calculated. b) The Cs climates appear in East Asia, especially along the Japan Sea coast of Japan. This type of climate differs somewhat from the original Mediterranean climate. c) There are transitional climatic areas in East Asia. Japan and its surround-

ings has an especially complex climate. Climatic conditions in these areas vary from year to year.

3) By calculating the frequency of distribution of climatic types that appear in the year climate, areas with stable and unstable climates were defined. Stable climatic areas are found in inland China, Indochina and in some equatorial regions. Unstable climatic areas are found in East Asia and India.

ACKNOWLEDGEMENTS

The writer is indebted to many people for this work. Prof. E. Fukui, Prof. T. Asai and Prof. M. M. Yoshino gave valuable advice and suggestions and Dr. T. Kawamura, Dr. I. Kayane and the other members of the present study group pointed out many problems and supplied much material. Students in the Geography Department at Mie University were helpful in calculating and drawing the maps used in this paper. The writer gratefully acknowledges their assistance.

References

Ackerman, E. A. 1941: The Köppen classification of climates in North America. *Geogr. Rev.* **31** 105–111.
Blüthgen, J. 1964: Allgemeine Klimageographie. Walter de Gruiter 599S.
Dobby, E. H. G. 1962: Monsoon Asia. A Systematic Regional Geography Vol. 5, Univ. of London Press 381p.
Finch, V. C., Trewartha, G. T. 1942: Elements of Geography. McGraw-Hill 823p.
Fukui, E. 1956: Revision of the Köppen climatic map of eastern Asia with new available materials. *Journal of Geography* **65** 149–158.
Fukui, E. 1961: The secular movement of the major climate areas of eastern Asia. *Geographical Studies presented to Prof. Taro Tsujimura in Honour of his 70th Birthday* 298–312.
Fukui, E. 1965: Secular shifting movement of the major climatic areas surrounding the North Pacific ocean. *Geogr. Rev. Japan.* **38** 323–342.
Geiger, R., Pohl, W. 1954: Eine neue Wandkarte der Klimagebiete der Erde. *Erdkunde* **8** 58–61.
Haurwitz, B., Austin, J. M. 1944: Climatology. McGraw-Hill 410p.
Köppen-Geiger 1961: Klima der Erde (map). Hermann Haak.
Nemoto, J. *et al.* 1959: Monsoons (Kisetsufū). Chijin Syokan 294p.
Rusell, R. J. 1932: Dry climates of the United States: II. Frequency of dry and desert years 1901–20. *Univ. of Calif. Publ. Geography* **5** 245–274.
Sekiguchi, T. 1951: On the year climate in Japan. *Geogr. Rev. Japan* **24** 175–185.
Stamp. L. D. 1961: A Regional Geography. Part IV. Asia Longmans 240p.
Strahler, A. N. 1969: Physical Geography. 3rd. ed. Wileys 733p.
Trewartha, G. T. 1954: An Introduction to Climate. McGraw-Hill 402p.
Trewartha, G. T. 1961: The Earth's Problem Climates. Univ. of Wisconsin Press 334p.

Materials

Arakawa, H. 1942: The Climate of East Asia (Daitōa no Kikō). Asahi Shimbunsha 189p.
Burma Meteorological Department: Climatological Summary (1948).
Clayton, H. H., Clayton, F. L. 1947: World Weather Records 1931–1940. *Smithsonian Miscellaneous Collections* Vol. 105. Smithsonian Institution.
English Meteorological Office 1958: Tables of Temperature, Relative Humidity and Precipitation for the World. Her Majesty's Stationary Office, London.
Flohn, H. 1958: Beiträge zur Klimakunde von Hochasien. *Erdkunde* **12** 294–308.
Fukui, E. 1942: The Climate of Southeast Asia (Nanpōken no Kikō). Tokyo-do 316p.

Hatakeyama, H. (ed.) 1964: The Climate of Asia. Tokyo-do 577p.

Malayan Meteorological Service: Summary of Observations (1946).

Meteorological Agency 1954: Annual Reports of Climate 679p.

Meteorological Agency 1967: Climatic Table for the World 113p.

Meteorological Agency 1969: Climatic Records (1951–1960). Technical Data Series (32) 181p.

Ogasawara, K. 1945: Climatic Problems in Southeast Asia (Nanpō Kikō Ron). Sansei-do 369p.

Philippine Weather Bureau: Annual Climatological Review (1952).

US Weather Bureau 1959: World Weather Records 1941–50. US Government Printing Office 1361p.

US Weather Bureau 1967: World Weather Records 1951–60. Vol. 4 Asia. US Government Printing Office 576p.

REPRESENTATION OF MOIST AND DRY CLIMATE OF MONSOON ASIA ACCORDING TO CLIMATIC INDICES

Takeshi KAWAMURA

Abstract: After a review of the climatic indices by Lang, Martonne, Köppen, Reichel, Emberger, Mayer, Prescott, Wang, Gorczynsky, Ångström, and Kira, four of them are applied to 550 stations in Monsoon Asia and their distributions are discussed. The patterns of distribution of these indices are similar, and it is proved that they are useful for representing moist and dry climate in Monsoon Asia. It is pointed out, however, that indices such as Lang's Regenfaktor, Martonne's aridity index and Ångström's coefficient of humidity should be revised for Monsoon Asia.

1. INTRODUCTION

The climate of Monsoon Asia is characterized by the seasonal alternation of wind systems. Generally speaking, while a dry climate covers almost the entire area in winter, a humid climate prevails in the same area in summer.

Representation of climate is divided into two categories, thermal regime and water economy. In this paper, the latter will be emphasized. In order to ascertain the degree of humidity or dryness of the climate, the most accurate method would be to measure or calculate the water balance of the climate in the area.

It is impossible, however, to measure evaporation and transpiration for a whole region directly. Therefore, the area amount of evapotranspiration is usually estimated by a combination of climatic elements, such as temperature, precipitation and so on. Most climatic indices are used to explain the climatic water balance.

Originally, such indices were mainly applied to the temperate region, especially in Europe. In Monsoon Asia, the collection of available data has not always been easy, and therefore the maps of climatic indices in previous studies are not always exact. The author has attempted in this study to provide detailed maps representing the moist and dry climate of Monsoon Asia using various climatic indices; Köppen's and Thornthwaite's indices are omitted, since they are treated elsewhere in this book.

2. CLIMATIC INDICES

Various climatic indices have been developed by climatologists. The first example of such an index is Lang's Regenfaktor (Lang, 1915). This index is represented as:

$$RF = P/T_{15}$$

where RF is the Lang's Regenfaktor, P the annual precipitation (mm), and T_{15} the mean temperature during the frostless period. As the calculation of T_{15} is troublesome, Lang improved this formula by replacing T_{15} with T_{20} (Lang, 1920). T_{20} is computed by the following equation:

$$T_{20} = \sum_{}^{n} t_{20}/12$$

where t_{20} is the monthly mean temperature above the freezing point. Lang pointed out that each type of natural vegetation classified, such as desert, semidesert, steppe and savanna, bush and forest, correspond to 0–20, 20–40, 40–60, 60–100, and 100–160 of the RF, respectively, and that an arid climate is below 40 of the RF, while a humid one is above 40.

Martonne (1926) improved this formula by substituting $(T+10)$ for T_{20}; this index is termed the aridity index (AI).

$$AI = P/(T+10)$$

where T is the mean annual temperature (°C) and P the annual precipitation (mm). According to this classification, the climate is arid when the value of AI is less than 20; values of AI below 5, 5–10 or 10–20 correspond to desert, interior drainage (grassland including steppe and savanna) and areas of dry agriculture, respectively.

Köppen (1936) devised a formula showing the relation between climatic elements and the timber line resulting in an arid climate, as follows:

$$AI' = P/(T+\alpha)$$

where α is a constant determined by the annual change of rainfall: i.e., $\alpha=7$ in an area where no dry season exists; $\alpha=14$ in an area where the dry season appears in winter; $\alpha=0$ in an area where the dry season occurs in summer. He pointed out that the isoline of $AI'=20$ coincided with the timber line.

Thereafter, Reichel (1928) revised the aridity index to the following form:

$$AI'' = P \cdot h/(T+10)180$$

where the correction term h is the number of days with precipitation and 180 is the mean value of h in Germany. Another revision was made by Emberger (1932):

$$Q = P/2\{(t_{max}+t_{min})/2(t_{max}-t_{min})\} \cdot 100$$

where t_{max} is the monthly mean value of the daily maximum temperature in the warmest month and t_{min} is the monthly mean value of the daily minimum temperature in the coldest month. Therefore, $(t_{max}+t_{min})/2$ is nearly equal to the annual mean temperature and $(t_{max}-t_{min})/2$ corresponds to the annual range of temperature.

The combination of temperature with precipitation as used here is based on the fundamental concept that the evaporation is closely related to temperature. Thus the ratio of precipitation to temperature corresponds to that of precipitation to evaporation.

Naturally, a number of attempts to use a climatic element more closely related to evaporation as a substitute for temperature were made over a period of years. For instance, the saturation deficit was taken by Mayer (1926), Haude (1955), and Prescott (1958). The simplest formula by Mayer is given as follow:

$$N/S \text{ ratio} = P/S$$

The improved ones are:

$$N-S \text{ ratio} = P/S^{0.75} \qquad \text{(Prescott, 1958)}$$

and

$$\text{humidity coefficient} = P/20 \sum_{s=1}^{n} S$$

However, the data available on saturation deficit are so sparse that it is impossible to draw a detailed map of the index of humidity. Thus, the much more complicated combination of temperature and precipitation had to be considered in order to represent a moist and dry climate.

Wang's aridity index (1941) shows the number of arid months, which are evaluated by a graph of which the ordinate and the abscissa show the monthly rainfall (p mm) and the monthly mean temperature ($t°C$) respectively. The curve of the hyperbola represented by the following formula is drawn:

$$p\{12p - 20(t + 7)\} = 3,000$$

If the plotted point of each climatic station on this graph is situated in the upper area of the hyperbolic curve the climate of the station is determined to be dry.

Other types of empirical expressions of climate were also found, for example, Gorczynsky's aridity factor:

$$A = \frac{1}{3} \operatorname{cosec} \varphi(Tr) \frac{P_{\max} - P_{\min}}{P}$$

where φ is the latitude, Tr the annual range of temperature, and $P_{\max}$ and $P_{\min}$ the maximum and minimum annual precipitation for fifty years, respectively.

Ångström's coefficient of humidity (AH) is shown by the following formula (Ångström, 1936).

$$AH = P/1.07^T$$

Kira's index of moist and dry climate (K) is calculated as follows (Kira, 1945):

$$K = P/(W + 20); \quad W < 100$$

or

$$K = 2P/(W + 140); \quad W > 100$$

where

$$W = \sum_{L=0}^{n} (t - 5).$$

This summation is carried out only on the months in which the monthly mean temperature exceeds 5°C.

Much more complicated combinations of the climatic elements were considered by Thornthwaite (1931, 1948) in order to represent the dry and humid climates. His theories on climatic representation are treated in detail by Kayane in another section of this book. His expressions, such as potential evapotranspiration and potential effectiveness, are excellent; however the basic problems have not yet been solved.

3. ARRANGEMENT OF DATA

The author intended to collect as much climatic data as possible for Monsoon Asia.

Table I. Number of available stations using climatic data.

Burma	12	Mongolia	4
Cambodia	6	Nepal	1
Ceylon	17	Pakistan	36
China	122	Philippine	16
India	105	Soviet	17
Indonesia	38	Thailand	14
Japan	64	Taiwan	14
Korea	26	Tibet and its surroundings	14
Laos	4	Vietnam	23
Malaysia	17		

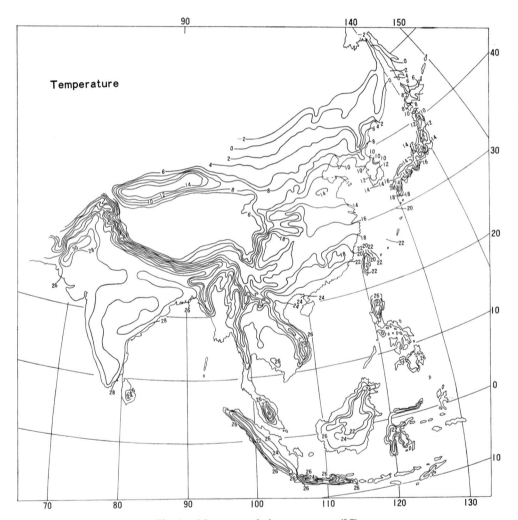

Fig. 1. Mean annual air temperature (°C).

However, available data on the direct measurement of moisture such as humidity, saturation deficit, evaporation, evapotranspiration, soil moisture etc. are so sparse that detailed distribution maps of climatic indices using such data could not be drawn.

Instead, climatic data on the monthly mean temperature and precipitation of some 550 stations were compiled. The number of available stations in each country is listed in Table I.

Climatic indices such as Lang's Regenfaktor, Martonne's aridity index, Ångström's coefficient of humidity and Kira's index of moist and dry climate were calculated for every station and detailed maps of these indices, the mean annual temperature and precipitation were delineated.

4. Results and Discussion

In order to consider water economy, the surplus and deficit of water should be estimated by the difference between rainfall and evapotranspiration. However, direct measurement of evapotranspiration from region is impossible, as mentioned above. Therefore the author first prepared a map of mean annual temperature (Fig. 1) and precipitation (see Kayane's paper in this volume). As it is well known that the temperature distribution is greatly affected by elevation, the isotherm is drawn taking the topographical conditions into consideration. Especially in areas where the observed data are extremely sparse, such as the Tibetan Plateau and the islands of Indonesia, this consideration is very important in preparing temperature maps. Of course, the isotherm cannot be drawn accurately in such a region.

Generally speaking, tropical Asia has a hot climate in the lowland areas, with a mean annual temperature above 26°C but not exceeding 28°C, except in some areas of India and Pakistan. The almost coastal plains in this region have similar temperatures. The climate drastically changes with the elevation in the inland areas of tropical Asia and the mean annual temperature is below 20°C at highland stations above 1,000 m.

On the other hand, the temperature gradually decreases with latitude in the temperate area of Monsoon Asia, including Japan, Korea, China, and southern Siberia. A particularly rapid change of temperature was found in northern China, northern Korea, and the northern part of Japan.

Although the precipitation regimes are discussed in detail in Kayane's paper, it may be said in summary that the influence of orography on the rainfall distribution is clearly perceived on precipitation maps. Rainfall is especially abundant on the windward side of the mountain ranges and scarce in the lee of such ranges. Monsoon Asia is known as one of the humid regions of the world; however, areas with relatively scarce rainfall extend not only inland in continental Asia, but also over the islands in the southern part of the Pacific Ocean. Therefore, the distribution of humid and dry regions can easily be estimated by the distribution of these elements. Distribution maps of the climatic indices are shown in Figs. 2–5.

a. Lang's Regenfaktor

In Fig. 2, the map shows an interesting feature of the distribution. The isoline 40 that divides the climate into two types, humid and dry, according to Lang's criteria, runs fairly smoothly from Mongolia to India. Small isolated pockets of dry climate are seen in the inland areas of Burma and Thailand. The other regions of dry climate are located along the eastern coast of India and in the eastern part of the East Indies.

According to Lang's definition, a dry climate includes desert and semi-desert, while a

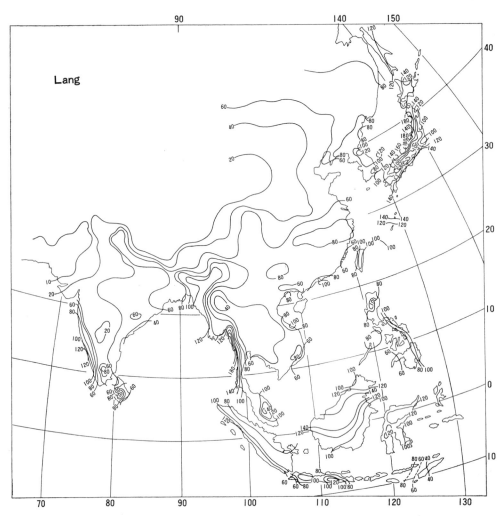

Fig. 2. Lang's Regenfaktor.

humid climate contains grassland, scrub and forest. If one compares this map with a vegeta-tion map, the area below isoline 40 in Fig. 2 includes monsoon dry woodland and thorn scrub in the tropical region. The desert and semi-desert region in the vegetation map cor-responds roughly to the area below isoline 20. On the other hand, the tropical rainforest, monsoon rainforest, broadleaved deciduous forest and taiga correspond to the areas where the Regenfaktor values are approximately above 80, 60, 50, and 70 respectively, though his paper indicated that these forest types are found in areas where the Regenfaktor values are between 100 and 160.

b. *Martonne's Aridity Index*

Figure 3 is much more detailed than the map drawn by Martonne himself, though the pattern of distribution of the former coincides roughly with that of the latter. According to

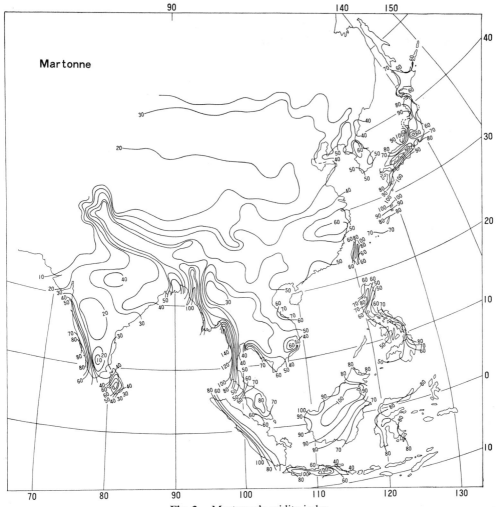

Fig. 3. Martonne's aridity index.

Martonne's map, the aridity index exceeds 40 in the inland areas of Burma, while most of the Korean Peninsula is below 40. Figure 3 shows the reverse, probably due to the difference in the data used.

Generally speaking, the distribution of Martonne's index is similar to that of Lang's Regenfaktor. Martonne's aridity index, however, is important in climatology and geography because of its close relation not only to vegetation but also to land use. In other words, the limit of cultivation caused by aridity corresponds to the values from 10 to 20 of the aridity index.

Yazawa (1963) studied the distribution of dry seasons throughout the world, using Wang's method, and compared this to the climatic arid boundary of cultivation. According to his results, the climatic arid boundary coincides with the isoline of 8 or 9 arid months in this region. This isoline corresponds to the 20 isoline of Martonne's aridity index. The forest

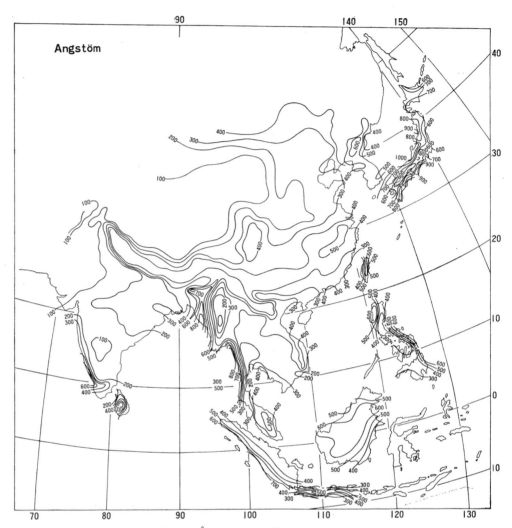

Fig. 4. Ångström's coefficient of humidity.

region is found in the area enclosed by the isoline of Martonne's index 30 in the tropical region, while the timberline runs through the region enclosed by the 20 and 30 isolines in temperate eastern Asia.

c. *Ångström's Coefficient of Humidity*

Ångström derived his coefficient of humidity from the empirical relation between the intensity of precipitation and temperature, based on observed data in Sweden, and mapped the distribution of this coefficient of humidity in northern Europe and its vicinity. He concluded that the coefficient of humidity has the following advantages:

1) At ordinary stations it seems to be closely proportional to the time of precipitation. 2) It runs closely parallel to Martonne's aridity index for monthly mean temperatures

Fig. 5. Kira's index of moist and dry climate.

between 0° and 20°C.   3) It is a continuous function of temperature and can be applied to investigations of arctic climates.

Consequently, his investigations originally excluded tropical climates with monthly mean temperatures above 20°C. Even in the tropics, however, the coefficient of humidity runs closely parallel to Martonne's aridity index in Fig. 3, as shown in Fig. 4. Nevertheless, the distribution of the absolute value of Ångström's coefficient of humidity tends to be larger in temperate climates than in tropical climates, in comparison with that of Martonne's aridity index. The forest limit caused by dry climate coincides fairly well with the isoline 200 of the humidity coefficient.

d.   *Kira's Index of Moist and Dry Climates* (humidity index)
Kira (1945) proposed a unique climatic representation corresponding to natural vegeta-

tion zones from the ecological point of view. In order to classify the climate, two indices were considered: the warmth index, or cumulative temperature during the growing season, and the humidity index or ratio between precipitation and the warmth index mentioned in the previous section. According to his criteria, climate can be classified into five climatic zones by the humidity index: humid, semi-humid, semi-arid, arid and super-arid. A humid climate corresponds to the following natural vegetation zones: tropical and subtropical rainforest, hard-leaved, laurel-leaved, broad-leaved deciduous, and evergreen forest. These forest types were classified according to the warmth index. Similarly, a semi-humid climate corresponds to summer green forest, temperate deciduous forest and so on. A semi-arid climate corresponds to savannah and needle-leaved deciduous forest and an arid climate to thorny scrub and steppe. A super-arid climate corresponds to desert.

The patterns of distribution of Kira's humidity index shown in Fig. 5 are similar to those of the natural vegetation, and the isoline 8, 6, 4, and 2 correspond to the boundaries dividing the humid, semi-humid, semi-arid, arid, and super-arid zones.

5. Conclusion

Detailed maps of four kinds of climatic indices for Monsoon Asia are shown in this paper. The patterns of distribution of these indices are similar, since they are based on the same categories; in other words, these classifications have been made empirically so as to fit the distribution of natural vegetation. These indices are useful for representing the moist and dry climate of Monsoon Asia. However, indices such as Lang's Regenfaktor, Martonne's aridity index and Ångström's coefficient of humidity should be changed to some extent if applied to Monsoon Asia, since they were primarily intended for Europe and other areas in the temperate zone.

New climatic indices should be investigated in future, taking into consideration the seasonal change of water balance.

Acknowledgments

The auther would like to thank Prof. M. M. Yoshino for his comments and encouragement, as well as Dr. M. Yoshitake, Director-General of JMA, and his staff for their kindness in assisting him in this study.

References

Ångström, A. 1936: A coefficient of humidity of general applicability. *Geografiska Ann.* **18** 245–254.
Emberger, L. 1932: Sur une formule climatique et ses applications à la botanique. *La Meteorol.* 423–432.
Gorczynski, W. 1943: Aridity factor and precipitation ratio and their relation to world climate. *Bull. Polish Inst. of Arts and Sci. in America.* No. 1.
Haude, W. 1955: Zur Bestimmung der Verdunstung auf möglichst einfache Weise. *Mitt. d. Deutch. Wett.* **2**(11) 1–24.
Kira, T. 1945: A new classification of climate of Asia as a basis of agricultural geography. Kyoto Univ. (in Japanese).
Köppen, W. 1963: Das geographische System der Klimate. *Handbuch der Klimatologie.* Bd. 1 Teil C. 1–44.
Lang, R. 1915: Versuch einer exakten Klassifikation der Böden in klimatischer und geologischer Hinsicht. *Inter. Mitt. f. Bodenkunde.*
Lang, R. 1920: Verwitterung und Bodenbildung als Einführung in die Bodenkunde. Stattgart 188.
Martonne, E. de. 1926: Une nouvelle fonction climatologique: L'indice d'aridite. *La Météorol.* 449–458.

Martonne, E. de. 1941: Nouvelle carte mondiale de l'indice d'aritide. *La Météorol.* 3–26.

Meyer, A. 1926: Über einige Zusammenhänge zwischen Klima und Boden in Europa. *Chemie d. Erde* **2.** 209–347.

Prescott, J. A. 1958: Climatic indices in relation to the water balance. Climatology and microclimatology. UNESCO 48–51.

Reichel, E. 1928: Der Trockenheitsindex, insbesondere für Deutschland. Tär-Ber. Preuss. Met. Inst., (162) 84–105.

Thornthwaite, C. W. 1931: The climates of North America according to a new classification. *Geogr. Rev.* **21** 633–655.

Thornthwaite, C. W. 1948: An approach toward a rational classification of climate. *Geogr. Rev.* **38** 55–94.

Wang, T. 1941: Die Dauer der ariden, humiden und nivalen Zeiten des Jahres in China. *Tübinger geogr. u. geol. Abh.* II (7) 33S.

Yazawa, T. 1963: Die Ariditätdauer und die kilmatische Trockengrenze des Ackerbaus in der Welt. *Japan J. Geol. Geogr.* **34** 211–216.

HYDROLOGICAL REGIONS IN MONSOON ASIA

Isamu KAYANE

Abstract: The distributions of water balance component, i.e., potential evapotranspiration, precipitation, storage of water and moisture deficiency and excess calculated by Thornthwaite's method for about 1,500 stations in Monsoon Asia, are presented in the accompanying maps and their characteristics are discussed. After comparing the precipitation with the potentional evapotranspiration, it was indicated that Monsoon Asia is drier than expected. Finally, the hydrological regions in Monsoon Asia are delineated in combination with the water balance components. Four main regions are distinguished: 1) A region of no annual water deficit. 2) A region of no annual water surplus. 3) A region in which both an annual water deficit and an annual water surplus occur. And 4) a region in which the annual water deficit is over 200 mm. Discussions on the irrigation problems in each hydrological region and the correlation between industrial activity and the lowering of ground water level in Tokyo are given.

1. INTRODUCTION

A number of attempts have been made by climatologists to classify the climate from the hydrological standpoint. According to Alissov *et al.* (1956), Voeikov had already classified climate into eight types as early as the 1880's on the basis of the river flow regime: "The river is the product of climate," he had stated. Since most living things cannot live without water, Woeikov's criterion was quite appropriate for delimiting geographical regions. Clearly, also, the river basin constitutes a hydrological system in which water circulation is controlled by geomorphic, climatic, biological, and other complex factors. As river discharge is a net output of the hydrological system, this implies a complex physical relationship between the heat and water budget of the basin. Unfortunately, however, Woeikov's classification, or variations of his system, have not been presented in map form in Asia yet.

Voeikov's idea was followed up by W. W. Dokucheyev in 1900 and the prevalence of a specified value for the ratio of precipitation to evaporation in different geographical zones was confirmed. G. N. Wyssozki had extended this idea in 1905, classifying European Russia using a series of indices similar to the ratio of total precipitation to evaporation (Alissov *et al.*, 1956).

In 1910, A. Penck developed a classification method in which the interrelation between precipitation-evaporation-runoff and groundwater was considered. The ratio of annual precipitation to evaporation, the phase of precipitation (solid or liquid), and the seasonal variation of precipitation were criteria of this classification and eight climatic sub-types were established (Blüthgen, 1964). Developing Penck's idea further, R. Lang in 1920 proposed the Regenfaktor, the ratio of the annual precipitation in mm to annual mean temperature in °C, to classify the soil; in 1926 Martonne (1926) developed his aridity index to define the boundary of interior drainage regions. The values of indices or quotients for delimiting a critical boundary of this kind, such as the Trockengrenze (Köppen, 1931) or *PE* Index (Thornthwaite, 1932) proposed after Martonne may be expressed by a common mathematical form:

$$P = a(T + b)$$

where P denotes the annual precipitation in mm, T expresses the annual mean temperature in °C, and a and b are constants depending on the seasonal precipitation and the boundary to be defined by the equation (Fukui, 1953).

Theoretically, evaporation data should be used to define the hydrological climatic condition, but many difficulties arise on observing actual evaporation from the land surface. The substitution of temperature for evaporation has been accepted in that the temperature shows a close correlation with the net radiation. Though the heat budget approach of estimating evaporation as made by Budyko (1956) or Penman (1948) is physically more sound than the empirical approach which uses temperature as a main variable, the latter approach is still valuable in revealing the main characteristics of a region where no homogeneous meteorological data are available except temperature and precipitation, as is the case throughout most of Asia.

2. HYDROLOGICAL REGION

We should first define what we mean by hydrological region. A hydrological region is a concept similar to a climatic region. Just as a climatic region is classified on the basis of similarities of climatic conditions, so a hydrological region may be classified by similarities of hydrological conditions. In classifying climatic regions, the selection of criteria to represent the climate is a major problem. Since climate is an abstract concept, unlike weather or meteorological phenomena, different criteria can be selected and different climatic classifications result. As a criterion for delimiting region, the hydrological condition is more realistic than the climatic one. Amounts of water surplus and water deficit and their seasonal variations are suitable for classifying the hydrologic region objectively. The surplus or deficit of water in a specified region is one of the important controlling factors for vegetation as well as human activity; thus the water balance problem is of both practical and academic interest. For a hydrological classification, the difficulty is not in the conceptual setting as is the case in climatic classification, but in finding an objective means of evaluating the water surplus or deficit in the region concerned.

The hydrological region may be defined by either of two methods. The first makes use of the river regime and the second of the climatic water balance. Other methods, which take into account the hydro-geomorphic or hydro-geological conditions of the region, are also valuable from a different standpoint, but they will not be discussed here.

The river regime method was first used by Voeikov, as stated earlier. The hydrological regions are delimited by characteristics of the river discharge, also called the river regime or *Abflussgang,* which was studied by European geographers. It is reasonable that studies of this sort should have been made in Europe where rivers are relatively important as transportation routes and the data on river discharge is relatively abundant. As examples of recent works along this line, we can point to Beckinsale's (1969) review paper on the river regime together with a paper by Grimm (1968) and one by Keller (1968) included in the "First Report of the IGU-Commission on the IHD" (Keller, ed., 1968). According to these papers, Lockermann (1957) also did research on the hydrology of Monsoon Asia for a dissertation in Bonn but this has not been published.

In its widest sense the river regime includes all occurrences and is delineated by a curve based on continuous or hourly observations (Beckinsale, 1969). In delimiting the hydrological region, the mean value, the mean of extremes and the absolute extreme for each month and for the year are commonly used. If sufficient discharge data are available, a detailed classification can be made. However, for global classification or for a large area, simple data such as the monthly mean are still not available.

The second method using the water balance approach has been greatly advanced by Thornthwaite's work on climatic water balance (Thornthwaite and Mather, 1955). The most useful aspect of this method is that no data other than temperature and precipitation are required. Since these are more readily available data than the river discharge, Thornthwaite's method is still applicable, though it involves some problems which will be discussed in the next section.

Thornthwaite's original concern when he defined the concept of potential evapotranspiration (*PE*) was climatic classification. The indices of humidity and aridity, as defined by the ratio of water surplus or water deficit to *PE*, and the moisture index defined by these two indices are criteria for climatic classification. The present author's concern is not in the climatic classification, but in the delimitation of hydrological regions. For this purpose using water surplus or water deficit data is more straightforward than using the indices stated above. Distribution maps of the water balance component are shown in Section 4, together with the hydrological regions that they define.

3. THORNTHWAITE'S METHOD OF WATER BALANCE

Two points should be mentioned with respect to Thornthwaite's method of water balance. The first is the accuracy of the *PE* value and the second is the method of calculating soil moisture storage. These two points are discussed here.

As an example of Thornthwaite's method of climatic water balance, the water balance table for Calicut, India is shown in Table I. Calicut is located on the west coast of India and the effect of the summer monsoon is apparent in the marked contrast between the wet and dry seasons in the year. Though the annual precipitation exceeds the annual *PE*, a

Table I. Water balance for Calicut, India (11° N, 76° E) in mm, 1881–1940.

	J	F	M	A	M	J	J	A	S	O	N	D	Y
PE	132	128	155	162	169	145	133	134	136	142	137	132	1,705
P	8	5	18	91	236	841	826	437	201	262	140	25	3,090
ST	138	91	57	44	111	300	300	300	300	300	300	209	
AE	79	52	52	104	169	145	133	134	136	142	137	116	1,399
D	53	76	103	58	0	0	0	0	0	0	0	16	306
S	0	0	0	0	0	507	693	303	65	120	3	0	1,691

PE: Potential Evapotranspiration   After Thornthwaite Associates (1963).
P: Precipitation
ST: Soil Moisture Storage
AE: Actual Evapotranspiration
D: Water Deficit
S: Water Surplus

water deficit occurs in the dry season, from December to August. As can be seen in Table I, the following relations exist for the annual totals:

$$P - AE = S = \text{Actual runoff}$$
$$-)\ \underline{PE - AE = D}$$
$$P - PE = S - D$$
$$= \text{Net runoff} \leq \text{Actual runoff}$$

The difference between the annual surplus and the annual deficit is therefore usually not zero. The difference between the annual precipitation and the actual evapotranspiration is the annual water surplus, and is equal to the actual annual runoff from a basin. However, a positive value of water suprlus for a basin does not necessarily mean a lack of water deficit in the basin. If all of the seasonal water deficit is completely offset by the construction of an irrigation system within the basin, the resulting runoff may be called "net runoff."

With respect to the first problem, it is necessary to think of the principle on which Thornthwaite's method is based. As stated earlier, its theoretical basis is that the temperature shows a close correlation with the net radiation which is the main energy source of evaporation. The grade of this correlation is, however, different for different climatic conditions. It is a limitation of Thornthwaite's method that his formulae are constructed on the basis of data observed mostly in the United States. The author has calculated the annual evapotranspiration loss from a basin in Japan using the watershed water balance method (Kayane, 1968); the annual evapotranspiration is calculated as the difference between the annual precipitation and the annual runoff. A value of 610 mm, which corresponds to about 80 % of Thornthwaite's PE of 750–800 mm, was obtained as a fourteen-year average. Taking into account the decreasing trend of the PE with increasing altitude in mountainous regions and the fact that the meteorological stations are mainly in valleys, the value of 80 % is probably realistic as a basin average. The relationship between annual precipitation— annual evapotranspiration—annual runoff for Japanese drainage basins ranging from a few square km to more than 5,000 square km in area are summarized in Fig. 1. Circles and dots in this figure represent individual drainage basins indicating that an annual evapotranspira-

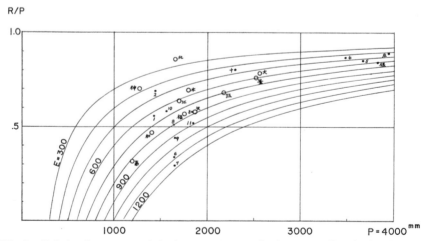

Fig. 1. Relation between precipitation - evapotranspiration - runoff ratio for Japanese drainage basins.

tion of about 500 to 800 mm is representative in Japan. A detailed explanation of the figure is given elsewhere (Kayane and Takeuchi, 1971).

As far as the annual total is concerned, Thornthwaite's *PE* value seems reasonable for Japan considering its location in the humid middle latitudes, as can be seen in Figs. 1 and 4. Though the annual total of Thornthwaite's *PE* is fairly reasonable, its seasonal distribution is not necessarily acceptable. Table II shows the *PE*'s for Tokyo calculated by Thornthwaite's and Penman's methods. Though the annual totals are very close, Thornthwaite's values are larger than Penman's from June to November, while the reverse is true for the rest of the year. The same systematic trend is also confirmed for Copenhagen by Aslyng (1960) and for Yuma, Arizona by Sellers (1964). The seasonal systematic difference between the *PE*'s observed and calculated by Thornthwaite's method is also discussed by Nieuwolt (1965) for Malaya. In the tropics, especially in the monsoon region, the seasonal variation of temperature is small whereas the variations in cloudiness, humidity, and precipitation are relatively large. This causes a systematic difference between the observed and calculated *PE* which Thornthwaite (1951) himself discussed. Further evaluation will be necessary when we study the water balance of different climates on a monthly basis.

Table II.   Potential evapotranspiration for Tokyo.

	J	F	M	A	M	J	J	A	S	O	N	D	Y
Penman	17	34	56	89	114	97	129	125	84	50	26	14	835
Thornthwaite	4	6	19	47	83	115	161	157	110	65	31	11	809
Difference	13	28	37	42	31	−18	−32	−32	−26	−15	−5	3	26

The second problem has already been discussed elsewhere by the author (Kayane, 1967). Since it is impossible to give the ratio of *AE/PE* as a unique function of soil moisture, any method can be used as a first approximation, as stated by Lowry (1959). Thornthwaite originally assumed, for convenience' sake, that the root zone of the soil contained a maximum of 100 mm of water in storage when fully moistened and that this moisture would be used at the potential rate as long as any of it remained. Later work (Thornthwaite and Mather, 1955) suggests that at least 300 mm depth of water is available for use by deep-rooted mature plants in most normal soils and that the evapotranspiration rate, which diminishes as the soil dries, is proportional to the amount of water in the soil. The water balance data used in the next section were calculated on this assumption. Since the amount of soil moisture available affects the values of *S* and *D*, any further discussion of the question without first considering the mechanism of soil moisture movement is futile.

4.   DISTRIBUTION OF THE WATER BALANCE COMPONENT IN MONSOON ASIA

A distribution map of the climatic water balance component is given in this section. Thornthwaite's method of water balance computation on a monthly basis has been applied to all regions in the world. The results have been published by C. W. Thornthwaite Associates (1963) as a series called "Average Climatic Water Balance Data of the Continents." The data used for drawing the distribution maps in this section were obtained from this source.

Thornthwaite's concept of *PE* has reportedly been applied to parts of Monsoon Asia such as Japan (Sekiguti, 1950; Fukui, 1957), Taiwan (Chen *et al.*, 1956), China (Tao, 1950; Chang, 1955; Chen *et al.*, 1956), India (Subrahmanyam, 1956), Pakistan and Burma

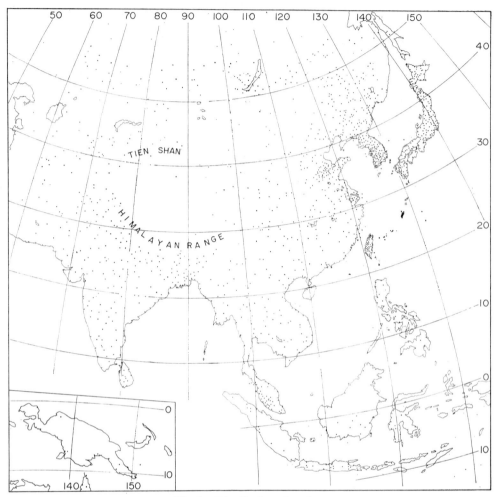

Fig. 2.   Distribution of stations.

(Shanbhag, 1953) but no distribution map of the water balance component is available for all of Monsoon Asia. One exception is found in Fukui's (1958) paper in which maps of *PE* and climatic region defined by Thornthwaite's new method (Thornthwaite, 1948) for all of Asia, including Siberia, Turkey, Arabia and Papua are presented. As the number of stations used was less than 150, however, detailed information for Monsoon Asia where the topographic effect on climate is marked is difficult to obtain from these maps.

Figure 2 shows the location of stations used for the present study. The total number of stations is about 1,500. As the geographical definition of Monsoon Asia is not clear, the region has been widened a little for convenience. Stations are dense in Japan, Korea, Taiwan, parts of China, and Malaya, moderate in India, Pakistan, mainland China, and Indochina, but sparse in Indonesia and the central part of the continent. The duration of the observation periods ranges from a few years to several decades. Original sources of meteorological data are listed in the "Aberage Climatic Water Balance Data." All stations included in this list are plotted in Fig. 2, except those that are duplicated or unknown. Many duplica-

Fig. 3. Distribution of annual precipitation. (base map: *Courtesy* Hammond Incorporated)

tions are found in China, perhaps because of the differences in spelling Chinese in English or German.

The distribution of annual precipitation in Monsoon Asia is shown in Fig. 3, whose base map showing the landscape features was originally drawn by A. K. Lobeck in 1945 and published by the Geographical Press, Columbia University. It is a well-known fact that a mountain range against a prevailing moist wind marks a clear climatic divide. This is especially true in the coastal region of Monsoon Asia. It is useful to make a physiographic map the base map for precipitation distribution, otherwise areas with heavy precipitation would be exaggerated in their horizontal extension, since the distribution of meterological stations is not dense enough to depict the climatic divide. In drawing the isohyetal line of Fig. 3 reference was made to the annual precipitation maps in "Climate of Asia" (Hatakeyama, 1964) for regions of sparse rainfall such as the southwestern part of Sumatra, the central part of Papua and so on. Papua Island is not shown in Fig. 3 because it lies outside the projection of the original base map.

Of the stations plotted in the figure, an annual total of 10,888 mm for Cherapunji, Assam,

(1906–1940) marks the maximum. Other heavy precipitation areas are found along the windward slope against the summer monsoon, and the sharp contrast to the leeward region is clear. This is true of Japan in winter where snowy weather prevails on the Japan Sea side and dry weather on the Pacific Ocean side, though this winter contrast becomes vague on the annual precipitation map. On the whole it may be said that Monsoon Asia is drier than expected. This fact will become clearer when we compare the precipitation map with the potential evapotranspiration map.

A distribution map of annual potential evapotranspiration is given in Fig. 4. The highest regions of over 1,800 mm are found in Thailand and southwest India at a latitude of about 10° to 15°N. If the topographic effect were fully taken into account, a more complex pattern would appear, like those constructed for Southwest Asia (Thornthwaite et al., 1955); however, the contour lines in Fig. 4 are already far more complex than previous maps (Fukui, 1958). It is interesting to compare Fig. 4 with the map of potential evaporation by Budyko

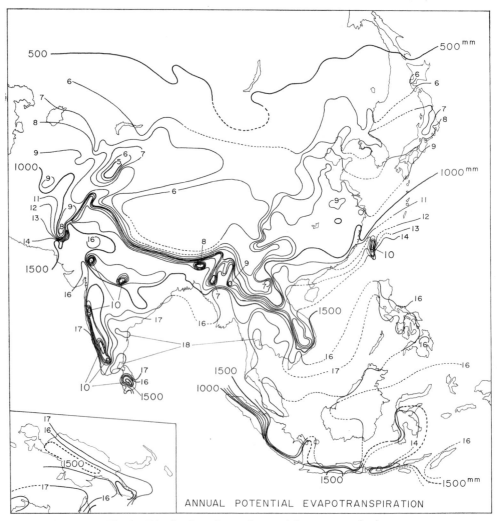

ANNUAL POTENTIAL EVAPOTRANSPIRATION

Fig. 4.   Distribution of annual potential evapotranspiration.

computed by the energy balance method (Academy of Science USSR, 1964). The greatest discrepancy, of more than 400 mm, is found in western India for which Budyko's map gives an evaporation potential of 2,000 to 2,250 mm. Under conditions of strong advection such as on an irrigated field in an arid region, Thornthwaite's method underestimates the true(?) value of potential evapotranspiration (Sellers, 1964). Though the potential evapotranspiration is defined as the moisture loss from the wet land surface homogeneously covered with short grass, such an ideal surface would not exists under natural condition. If a field is irrigated, then an additional heat supply by advection is inevitable. The relation between potential evapotranspiration and advection is something like a mirage.

Figure 5 shows the distribution of the annual water deficit calculated by the bookkeeping procedure, as shown in Table I. A similar figure was constructed for the annual water surplus but is not included here. As stated in the preceding section, a water deficit occurs at stations where the annual precipitation is larger than the annual *PE*. Therefore, the lines of

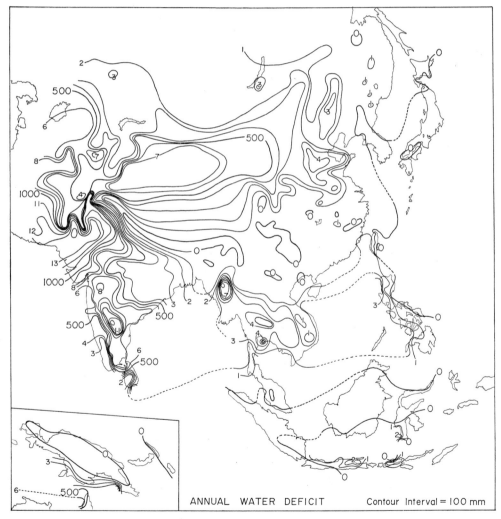

Fig. 5. Distribution of annual water deficit.

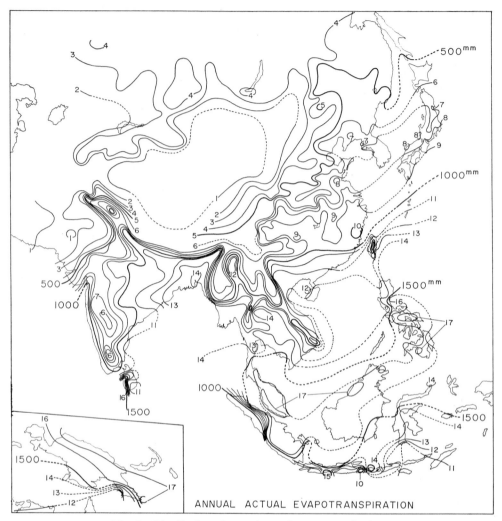

Fig. 6. Distribution of annual actual evapotranspiration.

no annual surplus and of no annual deficit do not coincide. It is interesting to note that the greater part of moist Monsoon Asia suffers from a water deficit in the dry season. The rainfall reliability of the respective regions makes the shortage of water even more serious.

Subtracting the values given in Fig. 5 from those given in Fig. 4, we obtain Fig. 6, which shows the distribution of the actual annual evapotranspiration. The maximum actual annual evapotranspiration over 1,700 mm is found in southwestern Malaya, east-central Borneo, northern Papua, and the central Philippines. Nieuwolt (1965) calculated the amount of evaporation in Malaya by various methods and made distribution maps of the monthly mean. According to these, 4.5–5.5 inches of monthly evaporation are expected along the southwestern coastal region of Malaya, amounting to 1,600 mm a year. In contrast, Budyko's map suggests a potential evaporation of 1,000 to 1,250 mm and an actual evapotranspiration of about 1,000 mm for the same region. The latter estimate seems a little low, though the true value is not known.

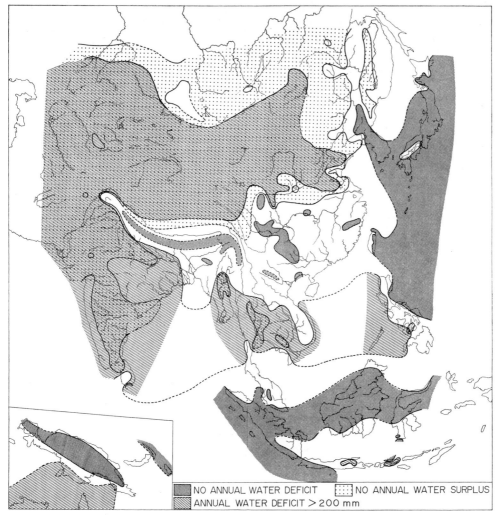

Fig. 7. Composite map of water balance components showing hydrological regions.

Combining these maps of water balance components, we can delimit the hydrological regions of Asia. One such attempt is shown in Fig. 7, in which the following five regions are distinguished:

A) Region of no annual water deficit (D=zero),

B) Region of no annual water surplus (S=zero),

C) Region in which both an annual water deficit and annual water surplus occur (both D and S are positive),

D) Region in which the annual water deficit is over 200 mm,

   D_1—without annual water surplus, and D_2—with annual water surplus.

We may of course define different hydrological regions by using different criteria. The division shown in Fig. 7 is based on the annual water availability in the region.

Curiously enough, the distribution of Region A is very limited. From agricultural standpoint it may be said that in all regions other than A there is one season in which irrigation is necessary. Dam construction for irrigation purposes is not effective in Region B where no annual actual runoff is expected. Region C is transitional zone where the possibility of using the seasonal water surplus for irrigation purposes exists. The 200 mm annual water deficit line has a somewhat special meaning, as will be discussed in the next section.

5. MAN'S ACTIVITIES UNDER DIFFERENT HYDROLOGICAL CONDITIONS

It is important to understand the relationship of human activities on the land to the natural environment. Environmental determinism has been blamed for attempting to explain cultural and economic phenomena deterministically by physical factors such as climate alone, but this does not mean that man can live in an artificial world. People living in some of the most actively deformed environments, such as Tokyo, are now confronted with the severe problem of Environmental Disruption (E. D.). The term E. D. (Kapp, 1970) was first used by a Japanese economist during a symposium on *Kôgai* (environmental pollution or disruption) held in Tokyo in 1970.

Nature has a tendency to maintain a state of balance or equilibrium by the process of self-regulation through circulating processes occurring near the surface of the earth. However, man's activities sometimes cut into this state of equilibrium, resulting in environmental disruption. Among the many E. D.'s occurring in severely industrialized regions, water pollution and land subsidence are closely correlated with the hydrological cycle. Water pollution occurs when the amount of industrial or municipal waste water discharged into a river exceeds its self-purifying capacity, which is a function of the flow rate and other chemical and biological properties of the river. Land subsidence occurs where artificial groundwater withdrawal from a groundwater basin exceeds the natural recharge. The circulation rate of groundwater is also governed by the hydrological condition of the region concerned.

One interesting example showing the close correlation between economic activities and

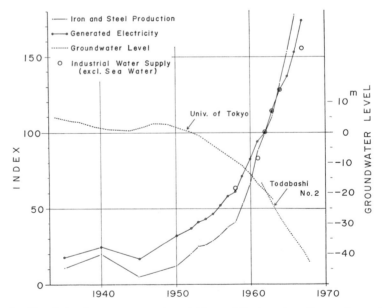

Fig. 8.   Correlation between economic activities and the lowering of groundwater level.

the lowering of the groundwater level in Tokyo is given in Fig. 8. The total production of steel and iron, the total amount of generated electricity, and the total amount of industrial water in Japan is expressed as a percentage of the values for 1962. Representing a close negative correlation to economic activities in Japan, the groundwater level in Tokyo was lowered by overpumping for industrial water. A detailed explanation of this correlation is given elsewhere (Kayane, 1970), but it is clear that the lowering of the water level has caused one of the most severe land subsidences in the world (Yamamoto and Kayane, 1970).

According to our experience in Japan, the amount of natural recharge for a groundwater basin composed of diluvium where the groundwater level is relatively deep is not more than 400 mm annually, whereas the withdrawal from the Tokyo area was more than 900 mm in 1963, at the maximum stage. Considering that natural recharge does not exceed 400 mm annually even in Region A, a value of 200 mm was selected as the critical value in Fig. 7. In other words in the monsoon region, a water deficit of 200 mm is taken as the upper limit which can be offset by groundwater recharged by the water surplus during the wet season. The 200 mm water deficit line indicates the boundary within which groundwater cannot be considered a renewable resource for offsetting the annual water deficit. It is interesting to note that in Fig. 7 the comparative positions of the lines of zero water surplus and 200 mm water deficit are reversed as we proceed to the higher latitudes. This means that in Region D_2, e.g. parts of India, Burma, and Thailand, the groundwater could be a renewable resource, but not in Region D_1 such as northern China and southern India.

Different means of water development and utilization are needed for different hydrological regions. These cannot be properly developed unless scientific knowledge on the hydrological cycle is made clear. The type of map shown in Fig. 7 should be useful for such purposes.

References

Academy of Sciences USSR. 1964: Fisiko-Geograficheskiy Atlas Mira. Moscow p. 22.

Alissov, B. P., Drozdov, O. A., Rubinstein, E. S. 1956: Lehrbuch der Klimatologie, VEB Deutscher Verlag der Wissenschaften. Berlin, 148–149.

Aslyng, H. C. 1960: Evaporation and radiation heat balance at the soil surface. *Arch. Met. Geophys. Biokl.* Ser. B **10** 359–375.

Beckinsale, R. P. 1969: River regimes. *In* Water, Earth and Man. Chorley, R. J. ed. Methuen London 455–471.

Blüthgen J. 1964: Allgemeine Klimageographie, Walter de Gruyter & Co., Berlin, 479–480.

Budyko, M. I. 1956: Teplovoi blans zemnoi poverkhnosti. Gidrometeorologicheskoe Izdatel'stvo, Leningrad (Japanese translation, 181pp.).

Chang, Jen-hu 1955: The climate of China according to the new Thornthwaite classification. *Annals Assn. Amer. Geogr.* **45** (4) 393–403.

Chen, C.-s., Sun, T.-h., Huang, T.-h. 1956: Climatic classification and climatic regions in China. *Res. Rep. Fumin Geogr.* Inst. Econ. Develop. Taipei, Taiwan.

Chen, C.-s., Huang, T.-h. 1956: The climatic classification and moisture belts in Taiwan. *ibid.*

Fukui, E. 1953: Physical geography II (Fukui, E. ed.), Asakura-shoten. Tokyo 22–25 (in Japanese).

Fukui, E. 1957: Thornthwaite's new climatic classification of Japan. *Tokyo Geogr. Papers I* 103–112 (in Japanese).

Fukui, E. 1958: New climatic classification of Asia according to Thornthwaite. *Tokyo Geogr. Papers II* 47–64 (in Japanese).

Grimm, F. 1968: Zur Typosierung des mittleren Abflussganges (Abflussregime). *Freiburger Geogr. Hefte* (6) 51–64.

Hatakeyama, H. 1964: The climate of Asia, Kokon-shoin, Tokyo, 577 pp. (in Japanese).

Kayane, I. 1967: Recent problems in evapotranspiration studies. *Tenki* (Bull JMS) **14** 271–284 (in Japanese).

Kayane, I. 1968: Variation in annual runoff ratio. *Freiburger Geogr. Hefte* (6) 25–32.

Kayane, I. 1971: Ground-water in Japan, *Mizu Keizai Nenpô* (Annual Rep. Water Economy), 23–65 (in Japanese).

Kayane, I., Takeuchi, A. 1971: On the annual runoff ratio in Japan. *Geogr. Rev. Japan* **44** 347-355.

Kapp, K. W. 1970: Can we continue to be alive? *Chuô-Koron*, May issue 148–157 (Original English title unknown, translated by K. Murata).

Keller, R. (ed). 1968: Run-off regimes and studies of the water balance, First Report of the IGU Commission on the IHD. *Freiburger Geogr. Hefte* (6) 240pp.

Keller, R. 1968: Die Regime der Flüsse der Erde. *Freiburger Geogr. Hefte* (6) 65–86.

Lockermann, F. W. 1957: Zur Flusshydrologie der Tropen und Monsunasien, Diss. Bonn, 619 pp. (unpublished).

Lowry, W. P. 1959: The falling rate phase of evaporative soil-moisture loss—A critical evaluation. *Bull. Amer. Met. Soc.* **40** 605–608.

Nieuwolt, S. 1965: Evaporation and water balances in Malaya *J. Trop. Geogr.* **20** 33–53.

Penman, H. L. 1948: Natural evaporation from open water, bare soil and grass. *Proc. Roy. Soc.* London **193** Ser. A 120–145.

Sekiguti, T. 1950:Climatological water balance problem in Japan. *Geophys. Mag.* **21**(2) 177–189.

Sellers, W. D. 1964: Potential evapotranspiration in arid region. *J. Appl. Met.* **3** 98–104.

Shanbhag, G. Y. 1953: Classification of the vegetation of India, Pakistan and Burma according to effective precipitation. *Roy. Inst. of Science* Doctor's Thesis Bombay **3** 1150pp.

Subrahmanyam, V. P. 1956: Climatic types of India according to the rational classification of Thornthwaite. *Ind. J. Met. Geophys.* **7** 253–264.

Tao, Shih-yen 1950: The moist and dry climate of China. *J. Chinese Geophys. Soc.* **2** 121–130.

Thornthwaite, C. W. 1951: The water balance in tropical climates. *Bull. Amer. Met. Soc.* **32** 166–173.

Thornthwaite, C. W., Mather, J. R. 1955: The water balance. *Publ. in Climatology* **8** (1) 86 pp.

Thornthwaite, C. W., Mather, J. R., Carter, D. B. 1958: Three water balance maps of Southwest Asia. *Publ. in Climatology* **11** (1) 57pp.

Thornthwaite Associates 1963: Average climatic water balance data of the continents. Part II and III. *Publ. in Climatology* **16** (1) 1–262, (2) 267–378.

Yamamoto, S., Kayane, I. *et al.* 1970: Simulation of groundwater balance as a basis of considering land subsidence. *Inter. Assn. Sci. Hydrol.* Publ. No. 88, 215-224.

Author Index

Subject Index

DSS

r

Y